Silica Stories

Christina De La Rocha · Daniel J. Conley

Silica Stories

Springer

Christina De La Rocha
Department of Geology
Lund University
Lund
Sweden

Daniel J. Conley
Department of Geology
Lund University
Lund
Sweden

ISBN 978-3-319-54053-5   ISBN 978-3-319-54054-2   (eBook)
DOI 10.1007/978-3-319-54054-2

Library of Congress Control Number: 2017932777

© Springer International Publishing AG 2017
This work is subject to copyright. All rights are reserved by the Publisher, whether the whole or part of the material is concerned, specifically the rights of translation, reprinting, reuse of illustrations, recitation, broadcasting, reproduction on microfilms or in any other physical way, and transmission or information storage and retrieval, electronic adaptation, computer software, or by similar or dissimilar methodology now known or hereafter developed.
The use of general descriptive names, registered names, trademarks, service marks, etc. in this publication does not imply, even in the absence of a specific statement, that such names are exempt from the relevant protective laws and regulations and therefore free for general use.
The publisher, the authors and the editors are safe to assume that the advice and information in this book are believed to be true and accurate at the date of publication. Neither the publisher nor the authors or the editors give a warranty, express or implied, with respect to the material contained herein or for any errors or omissions that may have been made. The publisher remains neutral with regard to jurisdictional claims in published maps and institutional affiliations.

Printed on acid-free paper

This Springer imprint is published by Springer Nature
The registered company is Springer International Publishing AG
The registered company address is: Gewerbestrasse 11, 6330 Cham, Switzerland

Preface

Two... four... six... eight...

In the most humble opinion of we two typical biogeochemists, everyone should have a favorite element. Ours is silicon. Did you just leap to a conclusion? Computer chips? Solar cells? They're neat, but not why we appreciate... silicon!

We heart silicon because it's at the heart of silica.

Silica is a compound, nominally SiO_2. Silicon dioxide you could call it, if you felt like sounding technical, or silicic acid, when it is dissolved in water. In any event, as molecules go, silica has versatility and a habit of participating in feats of derring-do.

Yes, feats of derring-do, and we don't mean the sci-fi dreams of silica-based life forms. What silica does in real life is cooler than a Horta. If you knew, you'd heart silica, too. That's what the next ten chapters, each telling a scientific tale of silica, are designed to do- make you fall in love with silica.

Silica was there, for example, when life began on Earth. In fact, it wasn't just there, it was key. It was the basis of the silicate rocks that reacted with hydrothermally warmed seawater to produce the solutes (dissolved substances) that reacted with each other and with metals to become the metabolic reactions at the core of all Terran life.

Much later on in geologic time, in the guise of stone tools made and used, silica helped to steer the evolution of human hands, cleverness, and ability to create and comprehend technology, music, mathematics, and complex compound sentences. This enabled us to become what we are today—intelligent and dexterous enough to have, for example, discovered, understood, and put to ubiquitous use in modern technology the piezoelectric properties of silica in the form of quartz crystals (if you press on them the right way, they give off electricity). Sonar, ultrasound, radios, telecommunications, you name it, it probably depends on quartz's piezoelectric effect.

During the billions of years in between the origin of metabolism and the invention of the quartz oscillator, some rudimentary animals, unicellular critters, and land plants developed the ability to biomineralize silica, producing microscopic scales, shells, skeletons, and liths featuring shapes, pores, lattices, grooves, spikes, and processes too fantastic to be matched by any mere human glassmaker (not even the venerable Blaschkas famous for their glass flowers and other equally incredible botanical and zoological glass models). More impressive yet, silica biomineralizers need no flame or furnace. They make their glass at room temperature or cooler. Incidentally, silica biomineralization is so common and widespread that, despite our best (unintentional) agricultural efforts to distill silica out of soils and into sewage systems, the world in your immediate vicinity teems with tons of microscopic, biomineralized glass.

Part of the reason that silica biomineralization is so common is that silica is everywhere. You can't throw a stone without hitting a silicate rock except maybe in the middle of a city (and even there, they're lurking immediately below the pavement, as the granite of fancy countertops, and on the outer walls of grander buildings). The water is full of dissolved silica, too, making it unsurprising that even vertebrate animals like ourselves, who don't biomineralize silica, have a true nutritional need for it. Our skeletons would be weak and malformed, our hair and skin would be a wretched mess, our organs would be falling apart, and our wounds would not heal if we hadn't kept up an adequate daily intake of dietary silica so far throughout our lives. Thank not only our water, but our beer and the plants we eat (especially grains). They're all full of easily digestibly absorbable silica.

All this silica cycling through lakes, rivers, the ocean, and the biosphere came from the dissolution of the silicate minerals that constitute silicate rocks, new ones of which are continually being produced through volcanism. The craziest thing about this slow dissolution of silicate minerals, which is known as chemical weathering, is not that it wipes out mountain belts (although it does), but that it is key to maintaining the pleasant, temperate habitability of Earth. When silicate minerals dissolve, they convert carbon dioxide from the atmosphere into carbonate salts that eventually end up in the ocean. In the longer term, this compensates for the steady release of carbon dioxide out of the magmatic interior of the Earth via such things as volcanoes and hot springs. Without silicate weathering, the surface of the Earth would have long ago baked itself sterile through a runaway greenhouse effect.

Silicate weathering will also mop up all the carbon dioxide we've spewed out via the burning of fossil fuels and forests too. Unfortunately for us, this will take more than 200,000 years. On the other hand, human beings are inveterate tinkerers. We have already found a way to speed up the chemical weathering of silicate minerals that, if it were to be deployed across broad swathes of the warm, wet tropics, could put some brakes on the global warming and climate change we've unintentionally unleashed.

Read on, please, for all the siliceous details.

Lund, Sweden　　　　　　　　　　　　　　　　　　　　　　Christina De La Rocha
　　　　　　　　　　　　　　　　　　　　　　　　　　　　　　Daniel J. Conley

Acknowledgements

Between the two of us, we (the authors of this book) have acquired 60 years of scientific experience of silica by working in limnology, oceanography, and biogeochemistry and by rubbing elbows with ecologists, microbial ecologists, paleoceanographers, and other Earth scientists. That's 60 years of learning *a lot* about silica from our own work and thinking and *even more* from that of hundreds of colleagues. It would take pages to name names. But, those of you who are reading this, you know who you are, especially those of you who have been especially special to us and the world of silica. We thank you for your passion for silica, for the diligence of your work, for the strength of your insights, for the excellence of your publications and presentations, for your guidance and generosity, and for the depth of the discussions we've had with you. Because of you, it's a great field to work in. Extra hats off go to those of you who read, commented on, and caught typos in early versions of the chapters and gave permission for figures to be printed in this book.

On a more practical level, this book could not have been written without the generous support from the Knut and Alice Wallenberg Foundation and from the Swedish Research Council (VR).

The friendly, stimulating, and coffee-, tea-, and cake-fueled environs of the Lund University Geology Department were no obstacle either. We thank all of you who are and were there for making it a great place.

Contents

1	**A Brief Introduction to the Players**	1
	1.1 Silicon	1
	1.2 Silica	2
	1.3 Silicic Acid	3
	1.4 Silicate	4
	1.5 Silicone	7
2	**The Origin of Life Was Brought to You in Part by Silicate Rocks**	9
	2.1 Setting the Stage	10
	2.2 A Flight of Fancy	13
	2.3 The Early Earth Was Not Hellacious	15
	2.4 A Fly in the Soup	17
	2.5 The Lost City	19
	2.6 Generating Organic Compounds	21
	2.7 Inventing Metabolism	23
	2.8 The World's Earliest Biological Carbon Fixation	24
	2.9 Replication	25
	Further Reading	27
3	**The Making of Humankind: Silica Lends a Hand (and Maybe a Brain)**	29
	3.1 Stone Tools and Their Makers	30
	3.1.1 The Earliest Stone Tools	31
	3.1.2 The Oldowan Industry and Its Practitioners	33
	3.1.3 The Acheulean Industry and Its Practitioners ...	35
	3.1.4 Neanderthals and the Levallois Technique	37
	3.1.5 Homo sapiens	39

	3.2	Hands and Brains	40
		3.2.1	Give Us a Hand	41
		3.2.2	If I Only Had a Brain	44
	Further Reading	47
4	**Mystical Crystals of Silica**	49
	4.1	What Is a Crystal?	49
	4.2	Pyroelectricity	56
	4.3	Piezoelectricity	58
	4.4	Sonar	61
	4.5	Quartz Oscillators	64
	4.6	But Why Is There a Piezoelectric Effect?	66
	Further Reading	67
5	**Glass Houses and Nanotechnology**	69
	5.1	Silica-Centric Musings on the Origin of Biomineralization	71
	5.2	The Early Fossil Record of Silica Biomineralization	74
	5.3	Not All Biomineralization Is Silica Biomineralization	76
	5.4	The World's First Arms Race	77
	5.5	How to Make a Glass House: Man Versus Nature	78
		5.5.1	Man	78
		5.5.2	Nature	80
	5.6	Some Silica Biomineralizing Organisms that We Are Learning From	82
		5.6.1	Choanoflagellates	82
		5.6.2	Siliceous Sponges	85
		5.6.3	Diatoms	87
	5.7	Siliceous Nanotechnology	91
	Further Reading	93
6	**Chicks Need Silica, Too**	95
	6.1	It's All About the Chicks	95
	6.2	Silicosis	97
	6.3	The Dog Days of Silica Medical Research	99
	6.4	Collagen	102
	6.5	Do Human Beings Require Silica?	104
	6.6	To Supplement or not to Supplement	108
	6.7	Silica, Aluminum, and Alzheimer's Disease	111
	Further Reading	113
7	**Of Fields, Phytoliths, and Sewage**	115
	7.1	All Plants Have Silica	116
	7.2	Opal Phytoliths	117
	7.3	The Benefits of Opal Phytoliths and of Dissolved Silica	120
	7.4	Is Silica an Essential Plant Nutrient?	122

Contents

7.5	Impact of Agriculture on the Silica Cycle	122
7.6	The Growing Creep of Silica Removal	124
7.7	Let's Go for a Walk Through Time	127
7.8	Silica in Sewage	130
7.9	A Plea for Hardy Souls	133
	Further Reading	133

8 Silica, Be Dammed! 135
   8.1 To Put It in a Nutshell 135
   8.2 A Brief History of Human Damming, or How Long
        Has This Been Going on 137
   8.3 Dams and Silica 139
   8.4 Dams, Eutrophication, and Silica 141
   8.5 Case Study #1: The Laurentian Great Lakes 142
   8.6 Case Study #2: The Baltic Sea 148
   8.7 Case Study #3: The Black Sea 153
   8.8 The Global View 155
   Further Reading 156

9 The Venerable Silica Cycle 157
   9.1 The Silica Cycle 157
   9.2 Silicate Weathering 159
   9.3 Getting Silica from Continent to Ocean 162
   9.4 The Weathering of Oceanic Crust 165
   9.5 Silica Biomineralization in the Ocean 168
   9.6 Silica's Return to the Mantle 169
   9.7 The Earth's Early Ocean Was a Tremendously
        Siliceous Place 171
   9.8 Silica, Cyanobacteria, and Banded Iron Formations 173
   9.9 And then Along Came True Silica Biomineralization 175
   Further Reading 176

10 Silica Saves the Day 177
   10.1 The Goldilocks Zone 178
   10.2 Most of Us Can Model 179
        10.2.1 The Warmth of the Sun 179
        10.2.2 Albedo, Which Is Not a Pasta Sauce 184
        10.2.3 Emissivity 186
   10.3 The Importance of Greenhouse Gases 188
   10.4 Silicate Weathering Consumes Carbon Dioxide 189
   10.5 The Temperature Dependence of Silicate Weathering 191
   10.6 The Paleocene-Eocene Thermal Maximum 193
   10.7 Enhanced Weathering 198
   Further Reading 200

Chapter 1
A Brief Introduction to the Players

Any undertaking on silica requires first setting a few terms straight, namely silicon, silica, silicic acid, silicate, and silicone. Even scientists get these words confused. So here it is stripped down to bare bone: *Silicon* is the element. *Silica* refers to a tetrahedron formed by one silicon atom bound to four oxygen atoms or to material consisting pretty much entirely of such tetrahedra. *Silicic acid* (also known as dissolved silica and dissolved silicate) consists of silica tetrahedra dissolved in water. *Silicate* is anything that contains silicon in a compound that acts as a negatively charged species (this means silica tetrahedra and silica tetrahedra where atoms of elements like aluminum have substituted in for some of the atoms of oxygen). *Silicone* covers a whole slew of silicon-containing organic compounds, generally man-made (and put to a variety of uses, such as lubricants, sealants, and novelty ice cube trays).

That quick rundown is enough to get you through the book, meaning you could skip straight from here to the first silica story (Chap. 2). But if you want to know a little bit more about these different silicon-containing materials and why they behave as they do, you should brave the rest of this chapter (which, by dealing with definitions, is the least exciting chapter of the book).

1.1 Silicon

Silicon is the name of the chemical element that sits on the periodic table between aluminum and phosphorus, underneath carbon, and above germanium, and at the center of an X formed by boron, nitrogen, gallium, and arsenic. All of these are fairly abundant elements, as is typical for the lighter end of the elemental spectrum. But silicon has most of them beat; it's the eighth most common element in the universe.

Silicon is even more common here on Earth, especially in the crust that we live on, where it is second in abundance only to oxygen. There's also no escaping silicon on Mercury, Venus, and Mars, on the Moon and the moons of Jupiter, or in roughly 96% of meteorites, either. You walk on silicon, breath it in, drink it up, and work with it all the time. In short, you've never spent a day of your life without being in contact with nontrivial amounts of it in some form or other.

1.2 Silica

But, despite its abundance, you've probably rarely encountered silicon on its own. Unless you are prone to making silicon wafers or computer chips (or smashing open electronic devices or solar cells), you could go your whole life without seeing silicon in its elemental form, which is a strangely lightweight, approximately silver-colored metal. In nature, such free, atomic silicon exists as the tinted darkness within smoky quartz and possibly not much else. Silicon prefers to be bound to oxygen, something there is no shortage of here on Earth.

Alone together, silicon and oxygen are known as silica. Under most surface Earth circumstances, the form they take is four oxygen atoms defining the corners of a triangular pyramid at whose heart sits a single atom of silicon, as sketched in Fig. 1.1 (where the dark lines represent the chemical bonds between silicon and the oxygen atoms and the dashed lines outline the pyramid). This silica tetrahedron forms the basis for most silicate minerals and it exists for two reasons. First, silicon, like carbon but unlike most other elements, has four bonds that it needs to fill with the help of other atoms. But, second, the four oxygen atoms that volunteer for the cause repel each other. The pyramid shape of the silica tetrahedron is the balance struck by the repulsive forces of the four oxygen atoms bound to the central silicon.

But a silica tetrahedron is more than just a stable chemical pyramid. It, if you can stomach the anthropomorphism, aches for more for even the four oxygen atoms leave a silica tetrahedron incomplete. Just as each silicon atom needs to engage

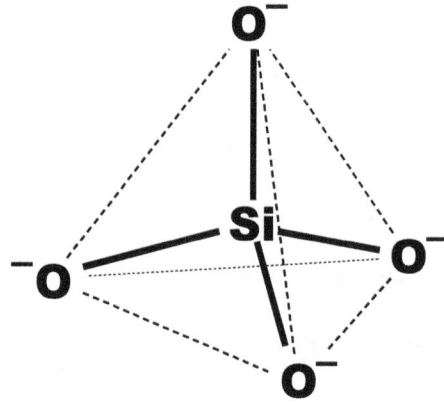

Fig. 1.1 The silica tetrahedron. The *solid lines* indicate bonds between atoms. The *dashed lines* outline the tetrahedral shape of the ensemble

1.2 Silica

Fig. 1.2 Silica tetrahedra joined to form silica

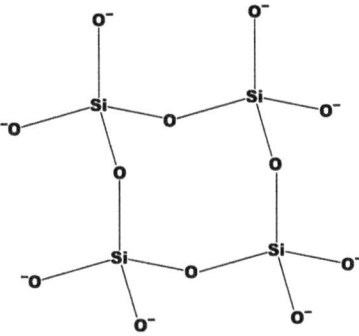

itself in four bonds, each oxygen atom needs to engage itself in two. This means each oxygen atom in a solitary silica tetrahedron is unsatisfied; none of the oxygen atoms in a solitary silica tetrahedron have more than their single bond with the central silicon. One solution to this is for a number of silica tetrahedra to come together so that each oxygen atom becomes shared by two silicon atoms, as in Fig. 1.2. Extraneous oxygen atoms are dispensed with until there are, on average, only two oxygen atoms for every one silicon atom. This yields a bigger, more complex silica, also known as silicon dioxide, also known as SiO_2.

"But!" you are saying and we hear you. No matter how many silica tetrahedra join together, unfulfilled oxygen atoms will remain at the periphery, as can also be seen in Fig. 1.2. In the presence of water, many of these terminal oxygen atoms pick up a hydrogen atom as their second bond, making for a number of OH groups (called hydroxyl groups) at the edges. That, finally, is enough to satisfy them.

1.3 Silicic Acid

Silica completely dissolved in water becomes a bunch of liberated silica tetrahedra whose oxygen atoms have each hooked up with a hydrogen atom (Fig. 1.3). This dissolved silica has the most fearsome name of silicic acid and its basic chemical formulation can be written as $Si(OH)_4$.

But silicic acid is not really fearsome. You drink it all the time and you've got it running through your veins (Yawn).

As with many acids, what makes silicic acid an acid isn't that it bores holes through things á la alien ectoplasm but that it can lose one or more of its hydrogen atoms to exist as a charged species (also known as an ion). In an alkaline solution, which is one that has a high pH and therefore a low concentration of hydrogen ions (H^+ in chemical notation), some of the $Si(OH)_4$ molecules let go of one hydrogen to become the silicate ion $SiO(OH)_3^-$ plus a free H^+. This addition of H^+ to the solution acidifies the solution by increasing its hydrogen ion concentration, lowering its pH. Thus is silicic acid an acid.

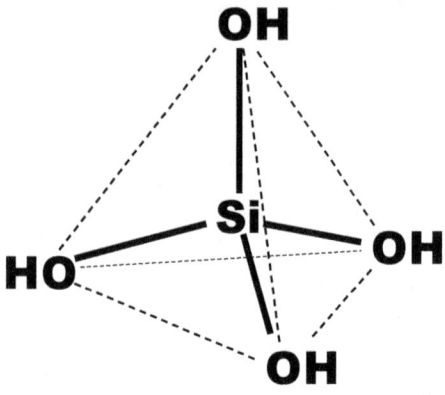

Fig. 1.3 Silicic acid, aka dissolved silica

In the book, we'll tend not to call silicic acid *silicic acid* but instead *dissolved silica*. Because we can. But also because it underscores the fact that dissolved silica is not a solid, but a solute.

1.4 Silicate

The silicates are where things start to get complex and we go a little bit nerdy on you. Bear with us!

Because of its four lonely, negatively charged oxygen atoms, the silica tetrahedron forms the basis for the wide variety of silicate minerals that make up the silicate rocks that constitute almost the entirety of the Earth's crust.

As pointed out in the brief rundown at the beginning of this chapter, silicates are materials that have silicon-containing units of negative charge. Anything that contains a silica tetrahedron fits this bill. This includes not only fairly pure silica like quartz and significantly more hydrated silica like opal, but also clays, serpentines, micas, feldspars, zeolites, garnets, zircons, pyroxenes, and amphiboles, just to name several mere broad categories of minerals.

These rock-forming silicate minerals can be incredibly different from one another in terms of their physical properties, chemical behavior, and appearance. Think of a wet clay, kaolinite, for instance. It has a smell, it can absorb water, it can dissolve in water, it is moldable, at least until it has been fired into a rigid form. Now think of a quartz crystal. It is transparent, hard, odorless, crystalline, and it takes not quite literally forever to dissolve in water. Both of these two minerals have silica tetrahedra as their basic basis. The difference between them is possible because of the chemistry of the silica tetrahedron, in large part due to the extra bond each corner oxygen desires. Depending on what ions are around to fulfill the oxygen atoms and on other factors, like the temperature and pressure at the time the mineral is forming, incredibly different interlocking (or not interlocking) mineral

1.4 Silicate

frameworks are possible. Silica tetrahedra may be linked to silica tetrahedra in chains, rings, or sheets, and these chains, rings, or sheets may be anchored together by regularly, repeatedly placed ions of other elements.

Without getting seriously into mineralogy, we'd like to give you a taste for this. While you could make it to the end of your days without knowing anything about this and suffer no ill effects, it is cool to pick up a rock and be able to imagine what lies within in terms of the crazy three-dimensional organization of its atoms. Plus, the underlying silicate structure of minerals is why they are what they are and they do what they do.

The simplest silicate structure consists of silica tetrahedra that are isolated from one another and thus not bonded together into a rigidly connected framework. To give one example, this is the case for the mineral olivine, which, when gemstone quality, is called peridot. In olivine, the disconnected silica tetrahedra are arranged in an orderly fashion within a sea of magnesium and/or iron atoms, depending on the type of olivine. A pure magnesium olivine, Mg_2SiO_4 is forsterite. A pure iron olivine, Fe_2SiO_4 is fayalite. And there are plenty of mixtures in between.

Believe it or not, the framework of olivine has global implications. Its loose, open structure makes it easy to dissolve and this dissolution, as with the dissolution of all silicate minerals, consumes carbon dioxide. Thus olivine's ease of dissolution makes it an ideal candidate for removing some of the excess carbon dioxide we've spewed into the atmosphere (a topic which is explored at the end of the final chapter of this book). This could help us avoid some of the global warming and climate change that is bearing down upon us, our descendants, and all other organisms on the planet.

The next step up is when silica tetrahedra bind together to form a regular, repeating, and predictable (that is, crystal) framework, such as in the case of quartz. But, visually speaking, the chains, double chains, rings, sheets of rings, and three-dimensional frameworks of rings of silica tetrahedra interbedded regularly with other ions are where things get interesting geometrically speaking.

You could go to any old mineralogy textbook or to any of dozens of websites to see the detailed molecular structures of specific silicate minerals (and you should; it's interesting). Plus the one thing this silica book isn't about is silicate mineralogy. So we decided to go the artistic route. Here's our pitch for what cool quilts, coloring books, or paintings silicate structures could inspire.

Take the cyclosilicates (Fig. 1.4). This family of minerals includes the beautifully blue, exceedingly rare, and correspondingly pricey benitoite (the very beautiful official gemstone of the state of California) and the beryl you may know as emeralds. In these cyclosilicate minerals, a regular number of silica tetrahedra link together to form rings (three tetrahedra in the case of benitoite; four, five, or even six in the case of other minerals, as you can see clearly in the geometric representation of beryl in Fig. 1.4). These rings of linked silica tetrahedra are isolated from the other rings of silica tetrahedra in the minerals and held in place within the crystal structure of the mineral by their bonds to other elements, hence the lovely regularity of the spacing between what looks like six-pointed stars in Fig. 1.4.

Fig. 1.4 An artistic representation of the structure of the cyclosilicate mineral beryl

Why does this happen? Those needy oxygen atoms. Because no matter how many silica tetrahedra you link together, there will always be some oxygen atoms that still need their second bond and this is where the other elements come in. In beryl, each six-tetrahedra ring consists of six silicon atoms and 18 oxygen atoms. This leaves 12 of the oxygen atoms in the ring lacking their desired second bond. The shortfall is filled by three beryllium atoms (which each seek to make two bonds) and two aluminum atoms (which each seek to make three).

This is all represented in Fig. 1.4. The aluminum atoms sit at the center of the rings of silica tetrahedra, as represented by the small, dark circles. The beryllium atoms sit between adjacent rings and, illustrated two-dimensionally, the bonds they have with neighboring rings are the diagonally bisected black squares anchoring the rings. The rings of six smaller sized white triangles at the surface of the image as it faces you are the rings of silica tetrahedra, as are the rings of six black triangles just beneath them. Unseen are all the rings beneath them. Who knew such complex tranquility could lurk within a crystal gemstone?

Sheet silicates, where rings of silica tetrahedra link directly together to form sheets of silica, are also pretty cool, although slightly less bedazzling (Fig. 1.5). In serpentine minerals (including the half-dozen referred to as asbestos), clay minerals (such as kaolinite, the main constituent of porcelain; vermiculite, friend to gardeners; and talc, friend to babies' bottoms but enemy to women's ovaries), and micas (including biotite and the sheety medieval Russian windowpane material muscovite), numerous flat sheets of networked rings of silica tetrahedra are

Fig. 1.5 A stylized representation of a sheet of linked rings of silica tetrahedra

1.4 Silicate

Fig. 1.6 A two-dimensional, suitable-for-quilting representation of the silica tetrahedra framework of a simple zeolite mineral

sandwiched between layers of things like aluminum and oxygen and are anchored by strategically located hydrogen atoms. What our simplified artistic rendering (Fig. 1.5) shows is a face on view of one sheet of networked rings of silica tetrahedra.

Zeolite minerals are also quite quiltable, although two-dimensional representations of zeolites, such as Fig. 1.6, fail to do them justice. Zeolite minerals, both naturally occurring and designed, consist of three-dimensional frameworks based on silica ring structures with connections on several layers. Their most noteworthy feature is spiraling stacks of rings of silica tetrahedra. These stacks are like microscopic tunnels through the mineral and can act like filters or traps for ions. They are what make zeolites so good at catalyzing chemical reactions and adsorbing pollutants. Industrially, zeolites are indispensable. But you probably know zeolites best as non-clumping cat litter; all those tunnels also make them great at absorbing liquid and entrapping malodorous entities like ammonia.

These and so many more are the ways that silica tetrahedra may be linked together with themselves and other ions to form all the different possible silicate minerals. But as this is a silica story book, not a silica coloring book, we will leave it here and hope two things: that you've picked up some appreciation for the behavior of the silica tetrahedron and that someone reading this will pick up the ball and run with it and start the art of silicate quilting.

1.5 Silicone

Before this chapter ends, we need to mention silicone. As a term, silicone covers a variety of polymers based on repeating units of Si–O–Si mixed in with atoms like carbon and hydrogen. Silicone crops up a lot in modern life, for instance, as Silly Putty and silicone caulk, grease, oven mitts, bakeware, and breast and testicle implants. There are silicone hydrogel contact lenses, as well, and silicone is common in shampoo and hair conditioner. With a list like this, silicone obviously has tales to tell. But silicone is not silica (meaning it is not made up of silica tetrahedra), so aside from a brief mention of what happens when silicones get into sewage, we won't be sharing its stories here.

Chapter 2
The Origin of Life Was Brought to You in Part by Silicate Rocks

Beginning at the beginning sounds like a good call. Or, if not at *the* beginning, at least at a big beginning. Will the origin of life on Earth do? Silica was more than there. Together silicate minerals, seawater, and hydrothermal heat set in motion a surprisingly simple sequence of chemical reactions that resulted in Earth's first metabolism, the first major hurdle overcome in the development of life.

Metabolism is a word that can make you think of the need to exercise and of those people, damn them, who can eat whatever they want to without piling on the pounds. But metabolism means much more than the rate at which calories are burned. Yes, the catabolic biochemical reactions of metabolism generate energy by oxidizing organic matter (made up of carbohydrates, lipids, and proteins). But metabolism also consists of anabolic biochemical reactions that consume energy during their construction of organic matter (namely the organic compounds that make up living organisms).

The initial invention of a metabolism capable of building things up as well as burning things down was a pretty big deal. Not only did it tick the first box of the instruction sheet *Basic Requirements for Life*, metabolic pathways tend to be long, convoluted chains of chemical reactions that start from a specific point and end at specific point, along the way producing specific products from specific ingredients.

Many metabolic pathways are complex enough to be the chemical equivalent of circular assembly lines. They take up a small compound and incorporate it into a larger molecule. They then either produce or consume energy while rearranging the larger molecule multiple times, as if solving a Rubik's cube. Eventually they break up the now rather rearranged larger molecule, eject the desired product, and rearrange the remaining materials to resurrect the enzymes and/or other molecules they started with. Then the cycle repeats.

As an example, take Fig. 2.1. That revolving wheel of chemical reactions, which is spinning continuously inside pretty much each and every one of your living cells, is the sort of thing a typical metabolic cycle entails. It is a complex masterpiece that managed to get itself invented out of chaos and disorder.

It would seem to be a miracle. But when you sit down and ponder the simple chemical reactions that would have been going on between water and silicate rocks

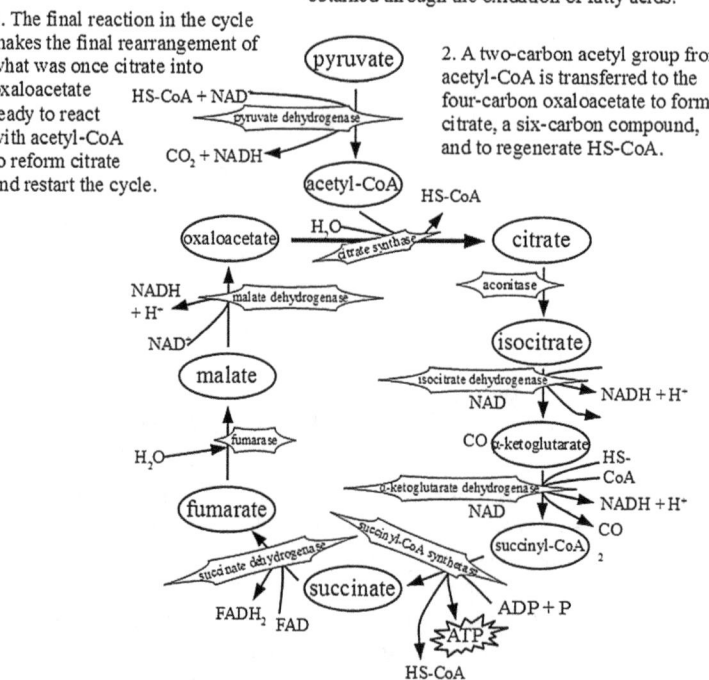

Fig. 2.1 All aerobic life on earth uses the citric acid cycle to generate energy by oxidizing acetate

on the flanks of undersea mountains approximately four billion years ago, the invention of metabolism starts to seem unavoidable.

And so we shall begin with this big beginning, or rather, a little bit before it because metabolism didn't invent itself in a vacuum. Metabolism came about on Earth shortly after the Earth's formation. The conditions that prevailed at that time set both requirements and constraints upon how life could have initially unfolded.

2.1 Setting the Stage

In the early days of our Solar System, there was no Earth nor any other planets, just a large, thin disk of matter rotating about the developing Sun. That disk consisted of gases and dust bumping, crashing, and crunching into each other. Sometimes when

2.1 Setting the Stage

things collided, they, or at least bits of them, got stuck together. Because bigger bodies are harder to smash to pieces than smaller ones, bigger bodies began to accumulate until the inner Solar System came to be a mess of smash-ups of perhaps half a thousand planetesimals on conflicting paths. Ka-blam! Ka-blam! Ka-blam!

Most meetings resulted in the larger subsuming the smaller. Each crash added to a to-be-planet's mass and continued to reduce the overall number of objects in the inner Solar System (sometimes $1 + 1 = 1$). Eventually, by this process, the crowded chaos of tiny bodies within the inner Solar System tidied itself up into the four terrestrial planets that are still there today (Mercury, Venus, Earth, and Mars) plus an asteroid belt.

The final smash for the almost-but-not-quite-Earth was with Theia. Theia, a proto-planet the size of Mars, was the smaller of the two. They may have been otherwise kindred, as surmised from the similarity in the composition of the present-day Earth and present-day Moon, which may be mostly made of Theia. If this is the case, it is likely that for millions of years, the proto-planet siblings had shared an orbit around the Sun. Accreting from the material that had accumulated at this same distance from the Sun would have meant the two protoplanets were made from materials of similar elemental and isotopic composition.

Locked in by the opposing pulls of the gravity of the Sun and the gravity of the proto-Earth, Theia either led or followed the proto-Earth by 60° as they orbited about the Sun.[1] The geometric side effect of this was that Theia circled the Earth (and vice versa) exactly one time for each full run around the Sun. So stable was this co-orbitation, Theia should still be there today, more than four and a half billion years down the line. (That would make for a very different story, one of life emerged on two similarly habitable planets and by now either learning to get along or fighting each other to death.)

But Theia is no longer there, just ahead of or just behind the Earth in orbit about the Sun. Thank the gravity of some passing planet, probably Jupiter, whose massive, wandering mass wreaked havoc in the Solar System in its early days. Tugged out of whack, Theia's orbit relative to the proto-Earth went from loopy but sustainable to simply dangerous. On one wobbly turn around the proto-Earth, Theia struck.

It's safe to say that none of us can really, fully grasp it, not even the modelers who study it using computer models and the laws of physics. It was no game of billiards. When one planetesimal struck the other, they didn't just briefly deform and ricochet. They demolished, shattered, melted, vaporized, fused, and ejected. These were the repercussions for the materials that made up Theia and the proto-Earth.

One long-standing theory for what then happened goes like this: The iron-rich core of Theia, too heavy to escape the gravity of the proto-Earth after impact, was trapped and sank to merge with the iron-rich core of the proto-Earth. To this day,

---

[1]If the Earth was at 6 o'clock on an almost perfectly circular orbit whose center was the Sun, Theia would have been at either 4 o'clock or 8 o'clock.

the hearts of the alien and the non-alien sibling remain intermingled deep below our feet. Meanwhile, the energy of the impact melted what wasn't already molten on the proto-Earth. Not just melted, in fact. An impressive portion of the planet vaporized and filled the atmosphere. At the same time, a substantial mass of material (mostly from Theia's mantle) flew off. Trapped in orbit about the Earth, it coalesced and became the Moon.

Despite the mass lost to the nascent Moon, overall the planet had grown by 10% as a result of the collision, which had, by the way, taken only hours. What remained when the crash was over had, at least for the first thousand postimpact years, a super-hot, silicate rock vapor atmosphere, but it was recognizable now as Earth with no front qualifier.

Sights upon the silver screen aside,[2] the crash with Theia was the last major collision that Earth has been involved in.[3] It is therefore the last incident in the accretion of Earth. You might call it the final moment of Earth's birth, or perhaps the last of Earth's many collisional, surface-melting, mass-adding rebirths. As such it marks the earliest moment that the clock could have started ticking on the aqueous geochemical reactions that resulted in life, for as soon as there was a permanently solid surface and some liquid water, conditions approached mild enough to host the precursors to life (and then life itself). More critically, conditions were finally stable enough for progress not to be necessarily totally wiped out in a subsequent, Earth-shattering catastrophe.

To know when this happened requires pinning the first point afterward that was cool enough for the rock vapor to have rained out of the sky and a crust to have formed over the magma ocean. In principle, one must just find and date a piece of this first crust. In reality, on Earth, there's none of this first crust left. It did not sufficiently survive its first few hundred million years of shattering by asteroid attack, much less the following billions of years of physical erosion, chemical dissolution, and, finally, tectonic subduction that recycles crust through the mantle to create new crust.

But all is not lost. This first crust that formed on the Moon fared comparatively well, for the Moon lacks wet weather, life, and plate tectonics. If we could figure out when this ancient lunar crust solidified, we'd have our answer.

Happily, aside from hundreds of thousands of asteroids, the most exciting things to have hit the Moon so far have been Apollo astronauts. In the true spirit of what it means to be human, they did not take only pictures and leave only footprints, they brought back almost half a ton of lunar rocks.[4] Much of what we know of the Moon and how and when it formed is based on the geochemical analysis of these rocks.

The dating of the formation of the Moon's crust using the oldest lunar rocks is a convoluted story, but here's the gist of it: A radioactive isotope of the element hafnium (hafnium-182), which was produced during the formation of the Solar

---

[2]We're thinking of Melancholia.
[3]The bolide that wiped out the dinosaurs? A fly upon the windshield.
[4]To be fair, we should point out that unmanned Soviet landers have also brought back lunar soil.

2.1 Setting the Stage   13

System, decays to a stable isotope of tungsten (tungsten-182). We know that there should be an excess of the daughter isotope (tungsten-182) in rocks from the Moon's surface because when the Moon solidified, the hafnium that was present would have concentrated itself in the Moon's silicate-rich mantle rather than distributing itself equally between the mantle and the Moon's metal-rich core. But some of the expected excess of tungsten in the lunar crust should be missing because the Moon formed millions of years after the formation of the Solar System, giving some of the radioactive hafnium time to decay away before the differentiation of the Moon into mantle and core.

Because we know that the half-life of the radioactive hafnium is nine million years, measurements of exactly how much tungsten is missing reveal that the Moon's crust formed 30 to 50 million years after the formation of the Solar System.

In other words, the Moon became cool enough to have a crust 4.53 billion years ago (give or take 10 million years, the error on the estimate).

Because the impact that formed the Moon is also the last event in the accretion of the Earth, 4.53 giga years ago must have been the end of the birth of the Earth and, again, now that the surface of Earth was not merely magma (or vaporized rock), the start of the clock for the geological chemical reactions that led to the bags of biochemistry called life.

2.2 A Flight of Fancy

Of course, life could have just flown in from outer space: microbes in meteorites from Mars (or Venus).

Although the surface and atmosphere of Mars currently lack vital signs, perhaps in the past, when the planet was young and pleasant, microbes teemed, lurking in the nooks and crannies of rocks and regolith. If so, what if, before Mars' internal heat expired and most of its atmosphere leaked to space, some of its life jumped ship?

For this possibility you might thank Jupiter, Saturn, Uranus, and/or Neptune. Their migration to their current orbits tossed 300 million years' worth of asteroids from the asteroid belt between Mars and Jupiter (or perhaps from the Kuiper belt just beyond Neptune) into the inner Solar System. What resulted was the Late Heavy Bombardment, which lasted from 4.1 to 3.8 billion years ago, and bombardment indeed it was. The scars can still be seen upon the helpfully non-regenerating surface of the Moon. Earth and Mars also took a beating. If there had been life on Mars at the time of the Late Heavy Bombardment, asteroid impacts would have launched microbe-infested rocks into space.

Some Martian meteoroids from that time are still out there, any hitchhikers within them long expired. (Billions of years is a long time to be cold, dry, and under vacuum.) But others made it to Earth, some only years after launch. Perhaps at least one deeply hibernating, safely rock-encapsulated microbe out of a Saganesque billions upon billions survived the trip and successfully colonized the Earth.

There are certainly meteorites that have been found on Earth that originated on Mars during the right time frame. Some of these meteorites contain microscopic structures and textures that look life-like. There are even scientists so sold on the idea, they grow bacteria in rocks that they then put under vacuum, freeze, and bombard with UV light before shooting them out of a cannon, across the lab, and into a wall to see if any bacteria survive. (Some do.) And let's face it, intrepid interplanetary bacteria have an undeniable charm. But this, as an explanation for how life began, passes the buck. Space traveling bacteria may indeed exist, and, although it dangles on a chain of ifs, this theory of interplanetary panspermia might explain how life initially arrived on Earth. But such a theory fails to explain how life initially came to be.

For the love of parsimony and of tackling questions directly, let us assume that life on Earth began on Earth. Perhaps, through experiment, observation, and consideration of thermodynamic and other constraints on chemical reactions, not to mention inspection of what scraps remain of Earth's most ancient geologic record, we can reason out a set of plausible steps.

Here is one thing we definitely know. By 3.8 billion years ago, photosynthesis, at least the more primitive anoxygenic kind, was widespread on Earth. Stable isotopes tell us so. The world's most important enzyme, Rubisco, captures carbon dioxide (CO_2) during photosynthesis so it can be used to make carbohydrate. Rubisco is better at capturing CO_2 containing the lighter stable isotope of carbon (carbon-12) than it is at capturing CO_2 containing the slightly heavier stable isotope of carbon (carbon-13). As a result, materials that include carbon that was fixed into organic matter during photosynthesis have a notably lower ratio of carbon-13 to carbon-12 than materials made from carbon that wasn't.[5] Such fingerprints of photosynthesis are all over the carbon in many of our oldest known rocks, such as the metamorphosed sedimentary rocks and the banded iron formations of the 3.6–3.7-billion-year-old Isua Greenstone Belt in southwestern Greenland and the 3.8-billion-year-old rocks of Akilia (also a part of Greenland).

These carbon isotope ratios screaming photosynthetic carbon fixation mean that by 3.8 billion years ago the process of life emerging from the chemicals of a lifeless landscape was long over. In addition, as neither the first life on Earth nor the last universal common ancestor of all life on Earth today (sweetly, LUCA, to those in the know) were photosynthetic, it also means that by 3.8 billion years ago we were already well past not just the first life form, but also well past the one form of life that gave rise to all the forms of life that exist today. Thus by 3.8 billion years ago, life had a firm foothold on Earth and was well into the (unintentional) task of improving itself through natural selection and, through the innovations that emerged, was busy increasing the capacity of the biosphere to capture energy, do

---

[5]The difference in the ratios is about 0.3%, something which is easy to measure using an isotope ratio mass spectrometer.

2.2 A Flight of Fancy 15

work, sustain biomass, and create ecosystems increasingly greater in size, complexity, and activity.

So that's the window we have to squeeze the origin of life on Earth into. Earth had a crust by 4.53 billion years ago and widespread and sophisticated photosynthesizers by 3.8 billion years ago. This window is less than seven hundred million years, especially as 3.8 billion years ago must considerably postdate the origin of life.

To some seven hundred million years is too short a span of time for this sort of thing. Life's kind of a big deal to invent and we don't mean that philosophically. There is no small number of disparate parts and processes, many of them bewilderingly complex, that have to come about, come together, and synergize before you can shout, "IT'S ALIVE!" There is an impulse to think that it took eons to pass this threshold.

But you can also ask why shouldn't life come about quickly when the circumstances are such that it would come about at all and you can say that seven hundred years is anyway hardly quickly. As a chunk of time, seven hundred million years is longer than the span for which metazoans (animals) have walked the Earth (or swum, or just sat there at the sediment–water interface). If seven hundred million years was more than enough time to get from brainless filter feeders to human consciousness, why shouldn't it have been more than enough time to get from geochemistry to biochemistry?

2.3 The Early Earth Was Not Hellacious

If life was widespread and way past LUCA by 3.8 billion years ago, it most likely originated soon after the formation of the Earth. Is that possible? Were those first few hundred million years following the crash with Theia not too violent and hot?

Researchers down through the decades have made many suggestions concerning conditions at the surface of the early Earth. Aside from being continuously attacked by asteroids, the Earth at that time might have been hot and dotted with lava for a long, long time and therefore not survivable by life until close to the 3.8 billion year mark. Or, in between the bolide blasts at least, the surface of Earth could have been colder than ice, with life (or the precursors of life) huddled around some deep maternal geothermal spring. And then there is the classic view of shallow seas of warm umami broth. Fire, ice, primordial soup. The possibilities span the gamut.

This first idea, that early Earth was hellish for quite some time after the violence of its formation, has fallen out of favor. For this you can blame a handful of crystals of zirconium silicate found in the Jack Hills of Western Australia. These zircons are up to 4.4 billion years old, which makes them the oldest objects of terrestrial origin we have found so far.

It is unlikely we will ever find anything much older. To look at, these zircons may not be impressive (none approach a millimeter in size), but they are durable. Wind, rain, oxygen, the temperatures and pressures of the geologic cooking process known as metamorphism,[6] no problems there. Zirconium silicate will outlast the other minerals in the rocks that formed through the cooling of the magma from which it crystallized. And outlast they have.

Once the Jack Hills zircons were liberated by the weathering away of the other minerals in their original rock, they were washed down slopes to form sediments with the remains of rocks also ancient but less so by several hundred million years. As time went by, these sediments became buried by further geological debris. More and more material accumulated on top of them. Eventually they were deep enough in the ground to experience the temperatures and pressures of metamorphism. This altered the sediments as a whole, deforming them on the macroscale like warm plastic. On the microscale, many minerals shifted their crystal structures due to the stress, while other minerals were driven to react with still yet others to form entirely different ones. But the zircon crystals remained relatively intact.

Over further eons, the material above the now metamorphosed sediments weathered away, finally exposing them and their zircons again to the light of day and, now that life has long overrun the Earth, the curiosity and destructive effects of geologists.

That these 4.4-billion-year-old zircons exist at all is an explosive fact. First, they require Earth surface temperatures cool enough for the creation of solid rock. Second, zircons form from the sorts of acidic magmas that yield not the basaltic rocks of the ocean floor, but the granitic rocks of the continents. Together that means that a mere one hundred and fifty million years after Theia blundered in, the Earth not only had a solid crust (which we had surmised already from the dating of the ancient moon rocks), it had produced a significant mass of the differentiated material that is continental crust.

Let's step back a bit and tell you that Earth has two types of crust, continental and oceanic. They both consist of silicate minerals, but that still leaves room for the two types of crust to significantly differ. Oceanic crust is similar in composition to the molten mantle that is its source—it is a magnesium- and iron-rich silicate comprising the sorts of rocks you see on Iceland or Hawaii. The silicate rocks of continental crust lean more toward aluminum, potassium, and sodium and are thus less dense. And so just as the mantle rides above the even more metal-rich (and therefore denser) core, continental crust floats higher above the mantle than oceanic crust and opens up the possibility for widespread areas of dry land surrounded by the sweeping expanse of deep, blue sea.

And that's the final explosive fact. The existence of these old zircons as components of a *sedimentary* rock means that the surface of the Earth those billions of years ago was not just cool enough for rocks and evolved enough to have two types

---

[6]Metamorphism refers to a change in the mineralogy of rocks at pressures and temperatures, generally around 150 °C (300 °F) to 850 °C (1560 °F), that are not high enough to melt the rocks into magma. Metamorphic effects include change in the crystal structure of the minerals or change in assemblages of minerals present in a rock, in both case towards those that are more stable at the temperatures and pressures involved.

2.3 The Early Earth Was Not Hellacious

of crust, it was cool enough for liquid water to exist and to weather rocks and wash the remnants down a slope to accumulate as sediment.

This means that from surprisingly early on, the Earth had a hydrologic cycle. And that means rain and runoff such as rivers across the continental landscape, even if the continental landscape back then was but a tiny fraction of the total of today's.

By the way, where there's liquid water, there is a possibility for life, and so we know that by at least 4.4 billion years ago, conditions on Earth were not hellacious, but for sure cool and wet enough to be survivable.

2.4 A Fly in the Soup

The idea that life assembled itself out of a so-called primordial soup, which can be traced back at least as far as Charles Darwin, inspired the classic experiment of Stanley Miller and Harold Urey, the one you probably heard about in your high school biology class. As experiments go, it must have been fun and not just because it was like playing god.

Miller and Urey set two big reservoir bulbs, Atmosphere and Ocean, into a circuit of glass tubing. Within this closed system they established what they believed to be the atmosphere on the early Earth: methane (CH_4), ammonia (NH_4), hydrogen (H_2), and water vapor (H_2O).[7] The heat source under the Ocean evaporated some of the water in that reservoir and a cooling jacket around a section of the tubing condensed the vapor to form rain that fell through the Atmosphere. But the best part of the experiment was the lightning; electrodes on either side of the apparatus discharged sparks through the Atmosphere. Their hypothesis was that this would spur the gases in the Atmosphere to react to form organic compounds[8] which would be collected by the falling rain and carried to the Ocean where they would accumulate. The idea behind this idea was that once you had organic compounds in your soup, they would react with each other, growing in variety and complexity until, voilà, life!

After merely a week, Miller and Urey sampled their Ocean and found they had succeeded, at least with the synthesis of organic compounds. There were lots in their Ocean and, most thrillingly, there was a broad suite of amino acids. (Amino acids, informally known as the building blocks of life, are not only what you need

---

[7]Today's atmosphere is 78% nitrogen (N_2), 21% oxygen (O_2), 0.9% argon (Ar), and 0.04% carbon dioxide (CO_2), and contains traces of neon (Ne), helium (He), methane (CH_4), krypton (Kr), and hydrogen (H_2).

[8]Organic compounds are molecules that contain carbon–hydrogen bonds and/or carbon–carbon bonds. Thus carbon dioxide (CO_2) which has only carbon–oxygen bonds is an inorganic compound despite containing carbon, while methane (CH_4), which has carbon–hydrogen bonds, and glucose ($C_6H_{12}O_6$), which has carbon–hydrogen and carbon–carbon bonds in addition to its carbon–oxygen bonds, are organic compounds.

to make proteins, they are used in the building of the building blocks of RNA and DNA.)

This was exciting and a major accomplishment. Miller and Urey had demonstrated both *that* and *how* organic molecules could be brought about by abiotic means. Even better, they'd shown that organic macromolecules (like amino acids) that are necessary for life and which were considered only producible by life, could in fact be abiotically generated under the conditions they thought representative of the early Earth. (Although, we'd like to point out, they had cheated by adding methane, an organic molecule, at the beginning of the experiment.) To many, this was one major chicken and egg problem solved.

Unfortunately, this primordial soup hypothesis still begs the question, *how did life begin?* You can make all the organic building blocks you want, but all the right ones still have to find each other in the dilute, crazy jumble that is the entire ocean and then order themselves into a fully functioning form before you have life.

One thing that life requires to do the things that make it life is enclosed spaces. These allow disequilibria to exist. Without disequilibria, life can't work (as introductory biology teachers are wont to shout: *Equilibrium is death!*). If a living cell were to find itself equilibrated with its environment or even fully equilibrated within itself, it would be unable to run any of its chemical reactions, pumps, or engines and that would be *game over*. For cells to do the things they need to do takes concentrations of solutes such as ions (and the charges associated with them) that differ between the inside of the cell and the outside, and between the inside of membrane-bound organelles inside the cell and the cytoplasm that fills the cell. In short, life requires chemical and/or electrical gradients in order to run.

Imbalances in charge or solute concentrations on either side of some barrier (such as a membrane) act as potential energy that is harnessed and used to accomplish various tasks. When ions move from an area of higher concentration to an area of lower concentration, they generate energy that the molecular machinery of the cell gathers and then releases in a controlled manner to drive work.

For example, if on one side of a membrane you have a lot of protons (those positively charged hydrogen ions, H^+), the protons will spontaneously slide through a membrane-spanning tunnel known as an ATP synthase. At the far end of this molecular tunnel of extremely specific construction, a molecule of adenosine diphosphate (ADP) and an atom of phosphorus tend to loosely attach themselves. When the proton, driven by the concentration gradient, passes through the ATP synthase, it drives a change its shape. The loosely attached ADP and phosphorus are pushed together, causing them to bind to form adenosine triphosphate (ATP).

This molecule ATP is super energetic and unstable. It exists solely to find somebody to take the third phosphorus off its hands. When it does, the energy that is released gets work done. If ATP hands the third phosphorus to a flagella, the flagella will beat. If ATP gives the energized phosphorus to a muscle cell, that cell will contract, helping to move the muscle. If ATP gives the phosphorus to a chemical reaction that needs a little kick, the reaction will go forward.

This work that must be done by cells in order to stay alive and fulfill their functions is all but endless—take up nutrients; break down molecules to release energy and raw materials; build new enzymes, carbohydrates, amino acids, DNA, etc.; pump out waste; grow and then divide in half; etc.,—and it is carried out by complex cellular structures or sets of structures that are specific, elaborate, and multitudinous. If one little detail about their size, placement, connectedness, or constitution is incorrect, they might not function at all and the cell will die. Because of this, it is unlikely that even the most minimally fully functional kit and kaboodle of life randomly assembled itself out of even a generous supply of lightning-generated liquid aminos even given eons. It's like that monkey typing Shakespeare. It could happen, but it won't.

Any reasonable theory for the origin of life needs to be less vague than "a lot of organic matter got made then somehow assembled itself into life." A reasonable theory for the origin of life needs to outline a set of feasible processes by which working mechanisms for catabolism (the gathering and controlled release of energy), carbon fixation (the production of organic molecules out of inorganic ones), biosynthesis (the construction of complex macromolecules from smaller organic components), and replication (cell division) came about and came to be incorporated together within a semi-permeable compartment (the lipid bilayer that is the cell membrane that all of Earth's truly living creatures possess).

Those are tall orders. But, we are happy to report, in recent years, serious progress has been made in the outlining plausible steps for the invention of catabolism, carbon fixation, and the basics of biosynthesis. This is where the silicate rocks finally come into the story (aside from them comprising most of the mass of the Earth).

2.5 The Lost City

It wasn't overall hellish hot, but at the time when life was somehow coming about in a non-monkey-typing-Shakespeare fashion, the Earth was still a rough place to be. The Late Heavy Bombardment of 4.1 to 3.8 billion years ago had to either bracket or postdate the emergence of life on Earth. Somehow early life got through it. It wouldn't have been easy. Just as small Solar System bodies in kamikaze mode were launching Martian rocks into space, ones large enough to vaporize the ocean and roast the surface of the planet were hitting the Earth with alarming frequency.

Thus there is another requirement for any theory of the origin of life on Earth. Life not only needed to make the series of leaps from inorganic matter to organic compounds to organized, enclosed, self-replicating, energy-controlling entities thriving by exploiting chemical disequilibria, it needed to be doing this in a place that wasn't being sterilized every century or so by the shock wave and blistering heat of an asteroid impact.

Geothermal springs have sprung to many a researchers mind. Like geysers and fumaroles (and volcanoes), they are places where energy from the super-hot interior of the Earth escapes in a concentrated way. On land you might know them as quiet hot springs under starry skies, in the gemstone hues of the Grand Prismatic Pool in

Yellowstone National Park and its rust-colored framing by microbial mats, or from the surreal blue pools of the travertine terraces of Pamukkale. In the ocean, hydrothermal vents and springs occur along mid-ocean ridges and are famous for sprouting ecosystems based not so much on the gutless and blind as on their chemosynthetic bacterial symbionts.[9]

Geothermal springs tick some serious boxes for kick-starting life. Sunk into the Earth as cracks and fissures, they provide protection from unpredictable extremes above (such as the oceans boiling off following an asteroid impact then recondensing). As water circulates through the Earth's crust near hot pockets of magma blurped up from below, it warms and it extracts some elements from the crust and loses some others that it has been carrying. When the now heavily solute-laden fluids exit the crust to mix with colder, more dilute groundwater, river water, or ocean water, they create environments whose main feature is disequilibrium. Cooling rapidly and mixing with waters more strongly oxygenated or of considerably differing acidity (or alkalinity), the geothermal waters find themselves unable to hold all of their solutes in solution. Minerals precipitate, producing chimneys that grow around the outflow. Pocked, fissured, and continuously generating, they are surfaces upon which reactions are catalyzed and, post-origin of life, microbes sequester themselves in highly productive biofilms.

In particular, it is the warm alkaline springs which exist on the peripheries of mid-ocean ridges that have captured the imaginations of those who work on the origin of life. Unlike their more famous cousins, the black smokers that belch billowing plumes of metal sulfides at 450 °C (840 °F), the warm alkaline springs, seeping fluids of roughly 100 °C (212 °F), are not too hot for life. In addition, their solute composition differs significantly from that of seawater (both most ancient and modern). When the warm alkaline waters mixed with seawater on the early Earth, they would have served as a factory for the sorts of engines life needs to carry out work.

Although such warm alkaline springs were probably common on the early Earth, their prime modern-day examples, the ones within the Lost City hydrothermal field, were not known to humanity until the year 2000. These Lost City seeps and springs are located at 30° 7′N, 42° 7′W, more or less midway between Jacksonville, Florida and the Canary Islands and just to the west of the rift zone running down the center of the Mid-Atlantic Ridge. They've sat there on the side of the Atlantis Massif for at least the last 120,000 years, which means that as geothermal systems go, they are long-lived.

The 4250-meter (14,000-foot) Atlantis Massif is itself a roughly two-million-year-old submarine mountain. Underneath its cover of sediment and scum, it must be beautiful. It is made of peridotite, a coarse-grained rock green with olivine and peppered with the black of pyroxene. This great lunk of silicate stone crystallized

---

[9]Chemosynthetic bacteria are like photosynthetic bacteria, except that instead of using solar energy to fuel the production of organic matter, chemosynthetic bacteria harness the power of chemical reactions.

2.5 The Lost City

from molten mantle below the oceanic crust. Later it was displaced and unearthed by the Mid-Atlantic Ridge during its continuing wedging apart of the eastern and western shores of the Atlantic Ocean.

When they were found, the Lost City springs, which terminate in tall towers of calcium carbonate, reminded us of something we'd known for a good long while: hydrothermal fluids flow through and react with silicate rocks like peridotite within the ocean floor in a process known as serpentinization. The Lost City springs also demonstrated that the resulting fluids can mix directly into seawater. And when you pay attention to the chemical reactions associated with the serpentinization of silicate rocks, you see a path that could lead to the carbon-fixing, biosynthesizing, and catabolic bases of cellular life.

2.6 Generating Organic Compounds

During serpentinization, seawater entrained into the crust and warmed reacts with peridotite. The olivine in the peridotite is converted into a mixture of two minerals, serpentine and ferrobrucite.[10] Because these warm fluids also contain abundant silica dissolved from the crust, the ferrobrucite reacts further to produce more serpentine, the mineral magnetite, and hydrogen in the form of H_2.[11]

Not to be unprofessional, but, *yow!* The early Earth didn't need to have Miller and Urey's conveniently (and probably incorrectly) hydrogen-filled atmosphere in order to make organic molecules. Hydrogen almost literally bubbles out of serpentinization. The right kind of silicate minerals, dissolved silica, and warm water (plus carbon dioxide) are all you need to start making organic matter and that is an incredibly big deal.

In other words, what happens once you have H_2? Organic compounds. To be more specific, the H_2 produced by serpentinization would have quickly reacted with carbon dioxide present in seawater to produce two organic compounds, formate ($HCOO^-$) and methane (CH_4),[12] no lightning, no life, and no cheating by having a magically already somehow H_2-rich atmosphere and an initial supply of methane.

The most brilliant thing about this proposed process for the abiotic generation of organic compounds? We don't have to do a lab experiment as a proof-of-concept. We know for a fact that these reactions happen out there in the ocean. Even today. The outflows from the Lost City warm springs contain ample abiogenic formate and

---

[10] $2(Mg_{1.8}Fe_{0.2}SiO_4) + 3(H_2O) \rightarrow Mg_{2.85}Fe_{0.15}Si_2O_5(OH)_4 + Mg_{0.75}Fe_{0.25}(OH)_2$
 i.e., olivine + water $\rightarrow$ serpentine + ferrobrucite.

[11] $14.25(Mg_{0.75}Fe_{0.25}(OH)_2) + 7.5(Si(OH)_4) \rightarrow 3.75(Mg_{2.85}Fe_{0.15}Si_2O_5(OH)_4) + 20.75(H_2O) + Fe_3O_4 + H_2$
 i.e., ferrobrucite + dissolved silica $\rightarrow$ serpentine + water + magnetite + hydrogen.

[12] $HCO_3^- + H_2 \rightarrow HCOO^- + H_2O$
 i.e., bicarbonate + hydrogen $\rightarrow$ formate ion + water.

abiogenic methane as well as other, larger, more complex organic compounds abiogenic in origin.

If you're reading this and wondering what the big deal is about generating organic compounds from inorganic materials without the involvement of life, here's what it is. As anyone who has suffered the slings and arrows of an organic chemistry laboratory course could tell you, organic compounds don't usually just come together. They're awfully hard to make. If you want to create an organic molecule from scratch, or even build a bigger organic molecule from smaller components, you have to know how much energy to put into drive the reaction, what chemical building blocks to begin with so that the right reaction occurs, and what catalysts to use so that you don't have to sit around and wait forever. Life is so darned good at this because it has immense arsenals of tools (ATP, NADPH, ion pumps, highly specific enzymes (i.e., catalysts), and scaffolding such as membranes) honed through billions of years of evolution for causing specific organic reactions to occur in a precise and controlled fashion.

Catalysts undoubtedly also prompted the abiogenic synthesis of formate on the early Earth (and still do today). The likely candidate is the nickel–iron mineral awaruite (Ni_3Fe). This mineral briefly binds dissolved carbon dioxide (in the form of bicarbonate ion, HCO_3^-) and hydrogen present in the water around or flowing past it. This brings the two molecules into close enough proximity to react with each other and become formate and water. It's a fairly run of the mill mechanism; by similar metal-catalyzed means are acetate (H_3CCOO^-)[13] and methane (CH_4)[14] formed via the addition of hydrogen atoms to dissolved carbon dioxide.

The biotic synthesis of formate, which goes on inside living organisms, also relies on a nickel–iron catalyst. This is likely no coincidence. Why would life invent a process from scratch when it could co-opt one already going on in the environment around it? Hydrothermal mounds, made of minerals precipitated from cooling hydrothermal fluids, abound with awaruite and other metallic minerals that catalyze such reactions. Even more interestingly, such metallic minerals often line the surfaces of exactly the sorts of microcavities that are shielded from the faster flow of fluids and therefore play host to chemical gradients that could be harnessed to get work done.

This is key. Catalytic mineral clusters dotting ragged surfaces within a flow of water can experience, establish, and maintain far-from-equilibrium conditions even without the aid of an organic membrane. In principle, if enough different catalysts were packed together in close proximity within the microcavities of growing geothermal surfaces, a suite of interacting reactions sufficient to be called *metabolism* could have developed prior to the invention of the cell membrane, a means of replication, or, indeed, any of the rest of the package of life. And that's the most

---

[13] $2(HCO_3^-) + 4(H_2) \rightarrow (H_3C)COO^- + OH^- + 3(H_2O)$
   i.e., bicarbonate + hydrogen → acetate + hydroxyl + water.
[14] $HCO_3^- + 4(H_2) \rightarrow CH_4 + OH^- + 2(H_2O)$
   i.e., biocarbonate ion + hydrogen → methane + hydroxyl ion + water.

2.6 Generating Organic Compounds

plausible scenario to be hit upon so far for the first step by which life began to be brought into existence on Earth.

It's also worth noting that the rough surfaces of hydrothermal mounds and chimneys become easily coated with abiogenic organic scum, making them not just good places to establish metal-catalyzed reactions, but also good places to invent organic membranes and easily jumble them up together with the catalysts, their chemical substrates, and their chains or cycles of reactions.

2.7 Inventing Metabolism

But we're getting ahead of ourselves. We've only gotten as far as methane, formate, acetate, and molecular hydrogen (H_2), and while that's within a hair's breadth of metabolism, it isn't quite there yet. Bear with us now as we start to really throw the letters around: not just C for carbon, H for hydrogen, and O for oxygen, but also S for sulfur, N for nitrogen, P for phosphorus, and a few other letters for a few other elements (mainly metals).

The methane, formate, acetate, and hydrogen being produced hydrothermally 4.4 billion years ago would have reacted, at times with the help of hydrothermal metal mineral catalysts, with the carbon dioxide, nitrate (NO_3^-), nitrite (NO_2^-), sulfite (SO_3^{2-}), sulfur (S), iron (Fe^{3+}), manganese (Mn^{4+}), and phosphate (e.g., HPO_4^{2-}) that existed in the early, not yet oxic ocean in series of steps to form some very key organic compounds. For instance, the green rust mineral fougèrite would have catalyzed the partial oxidation of methane by nitrate to form methanol[15] (that type of alcohol you shouldn't drink unless you would like to go blind or maybe die). Oxidation of this methanol by nitrite would result in formaldehyde.[16] Subsequent reduction, thiolation,[17] and condensation of the formaldehyde would easily yield thioacetic acid (H_3C–CO–SH). So far so simple and occurring over the sorts of spans of time that occur in a common chemistry laboratory even when it's filled with first year university students who are bumbling around.

The next step toward metabolism would be thioesters (which are any organic compound containing a C–CO–S–C functional group). There are numerous plausible ways they could have been produced under mild conditions. For instance, when prompted by catalysts of iron sulfide or nickel sulfide, warm water containing carbon dioxide, carbon monoxide, and hydrogen sulfide form methylsulfide (H_3CSH) which further reacts to form methylthioacetate (H_3C–CO–S–CH_3). As these organosulfur compounds reacted further in the alkaline springs, the thiol

---

[15] $CH_4 + NO_3^- \rightarrow (H_3C)OH + NO_2^-$
  i.e., methane + nitrate → methanol + nitrite
  This reaction couples the partial oxidation of methane with nitrate reduction.
[16] $(H_3C)OH + NO_2^- \rightarrow (H_2C)(OH)_2 + NO^-$
  i.e., methanol + nitrite → formaldehyde + nitric oxide.
[17] Addition of a sulfur that is bound to a hydrogen, i.e., a sulfhydryl group (SH).

known famously to biology as coenzyme A (of overall formula $C_{21}H_{36}N_7O_{16}P_3S$) would have started to accumulate. From there, further reaction with carbon dioxide and hydrogen to form acetyl-CoA ($C_{23}H_{38}N_7O_{17}P_3S$), which is just coenzyme A with an acetyl group (CH_3CO) joined to it at one end, would have been energetically inevitable.

What the world had now was the basis for cyclic metabolism. Not only is coenzyme A good at picking up an acetyl group to become acetyl-CoA, acetyl-CoA is very good at passing the acetyl group on to another molecule in a reaction that releases energy (the basis for catabolism) or builds up another organic molecule (the basis for anabolism). Further, in the process of losing the acetyl group, acetyl-CoA recycles itself back to being a molecule of coenzyme A capable of doing it all over again. That's the cyclic part.

So fancy that. Warmly weather the right silicate rocks at the bottom of the ancient ocean and within a short span of time and without any instructions encoded in DNA or provided by RNA, you'll end up with an abundance of two of the most important molecules of Terran metabolism.

Today we're billions of years down the line from the origin of life, but acetyl-CoA and coenzyme A still reside at the heart of how every single aerobic organism turns organic matter into energy in order to function and stay alive. There isn't, for example, a fully functional aerobic cell on Earth or inside of you that doesn't use acetyl-CoA to deliver carbon to the biochemical cycle known as the citric acid cycle (the one diagrammed in Fig. 2.1) and coenzyme A in a number of the steps subsequently needed to produce that mobile packet of energy known as ATP (without whose constant production you would quickly die because, proximately, it is the fuel that your body burns to do the work that it must do).

This thought is well worth repeating and so we will. Without any prompting, fundamental pieces of a functioning cyclical metabolism could have been derived from the byproducts of silicate weathering in a period of time too short to register even as a blip on a blip in Earth history.

2.8 The World's Earliest Biological Carbon Fixation

Conveniently and most interestingly for our hypothesis that ultimately serpentinization-derived coenzyme A is where it all began, coenzyme A and acetyl-CoA can also be used to fix carbon. To fix carbon means to take an inorganic form of carbon, such as carbon dioxide, and turn it into an organic form, perhaps, for example, a sugar or other carbohydrate. Photosynthesis is a type (but not the only type) of carbon fixation and there would be no big beautiful biosphere like the one we have today if bountiful amounts of biological carbon fixation did not occur.

2.8 The World's Earliest Biological Carbon Fixation

The two most ancestral forms of life that still exist on Earth, the methane-producing methanogens (which are archaea) and the acetate-producing acetogens (which are bacteria),[18] both fix inorganic carbon into organic matter using the metabolic pathway known as the reductive acetyl-CoA pathway (which is also known as the Wood–Ljungdahl pathway). This is a noncyclic pathway based on the formation of acetyl-CoA from coenzyme A much as described occurring in the warm, alkaline springs. Archaea and bacteria, however, speed up the reaction using enzymes.

During this reductive acetyl-CoA carbon fixation pathway as practiced by some modern day, if metabolically antiquated microbes, carbon monoxide (CO) is produced from carbon dioxide (CO_2) by throwing the extra oxygen atoms to hydrogen atoms to form water. This reaction is catalyzed by the enzyme carbon monoxide dehydrogenase (which uses our geothermally originated friend, a Ni–Fe reaction center, to bind the CO_2). With the help of another enzyme, the carbon monoxide reacts with a methyl cation (CH_3^+) to form acetate which then binds to coenzyme A to form acetyl-CoA. This acetyl-CoA, thus containing some carbon that was just fixed into it from an inorganic source, is then shunted into cellular biochemical pathways and consumed (instead of being regenerated) in the synthesis of other organic molecules. Alternatively, acetyl-CoA may be cleaved to yield acetyl phosphate which may easily phosphorylate ADP to yield ATP, the energetic molecule that is the proximate source of energy for all work, physical and chemical, done within living organisms (including you).

This is as simple as biological carbon fixation gets. It is also the reductive acetyl-CoA pathway that is the catabolic pathway that LUCA, that last universal common ancestor of ours, and its forbears would have used to create organic matter for themselves and their energetic needs. Essentially, once the self-sustaining generation of coenzyme A occurred abiotically, it was there to be co-opted into biosynthetic reactions and more complex metabolic pathways.

The evidence that this pathway, or some rudiment thereof, was the pathway all subsequent carbon fixation pathways evolved from is the fact that acetyl-CoA is a product of all currently known carbon fixation pathways, no matter how complex. Equally importantly, there are no contradictions. There is nothing about the proposed sequence that could not have occurred given what we know about the early Earth at the time when life needed to be coming about.

2.9 Replication

Summed up, it is highly plausible that in but a few fairly simple steps the hydrothermal weathering of silicate rocks in the deep sea of the early Earth gave rise to a carbon-fixing, energy-generating metabolism based on acetyl-CoA. For a

---

[18]Phylogenetically, the tree of life branches into three main domains, those of the Archaea, the Bacteria, and the Eukaryotes. We are Eukaryotes.

while, this metabolism occurred within the catalyst-lined microcavities at the outflows of warm and alkaline geothermal springs at the bottom of the ocean. As such, it was like life, but not quite yet life. There were still a few hurdles to overcome, mainly those of reproduction and of liberation.

Liberation is easier to imagine. Those microcavities, which tended to accumulate organic scum, just needed to become covered with polar lipids of the type that assemble into the lipid bilayer that encapsulates living cells. Then the metal catalysts and acetyl-CoA reactions could have become incorporated into a proto-cell. Proto-cells would have then sloughed off the walls like mad once enough acetyl-CoA, coenzyme A, polar lipids, and metal catalysts were being produced at a high rate. But this was still not quite life. No matter how numerous, nor how fast their rate of *production*, before such proto-cells could be pronounced life, they needed to possess a means of *reproduction*. They needed to be able to make new copies of themselves.

But we may have just gotten the order wrong. It could be that the problem of replication was solved before metabolism enclosed itself within sets of organic membranes. If this is correct, LUCA, the last common ancestor of all of Earth's currently living things, was not enclosed within an organic membrane. It would have still been living upon the surface of a microcavity when it developed the capability to reproduce itself.

A plausible, precise, and detailed hypothesis for the invention of reproduction remains the toughest nut to crack to finish the puzzle of the origin of life on Earth. To make a fully functioning copy of yourself, even if all you are is a simple cell with a cyclic metabolism, requires coding the instructions for building and operating yourself in a form that can be read and acted upon by molecules. Otherwise you're stuck on the wall of the warm spring and when its heat source dies out, no more of you will be produced. For most of life's tenure on Earth the instructions necessary to create and run a cell (or a multicellular organism) have been based on DNA, RNA, and tons of genes and lots of enzymes and messengers acting together. It is far beyond our current grasp the exact steps by which such an exquisitely functional tangle of complexity first came to be.

But we do at least have some vague ideas. The reactions in the warm, alkaline springs that led to acetyl-CoA and its friends would have fairly feasibly led further to the nucleic acid building blocks of RNA plus the ribose needed to hold its backbone together. Eventually the RNA led to DNA and with it came the possibility to set, read, and copy a set of instructions for the molecular machinery and engines needed to carry out the chemical reactions of a basic, functioning cell. That's more than likely the part of the origin of life that took the longest time by orders and orders of magnitude. You can almost laugh at cyclic metabolism and carbon fixation. What cakewalks. They're pathetically simple by comparison.

It was once there was DNA and the means of using and copying it that there was LUCA, again probably still in its hydrothermal mound microcavity and not yet enclosed by an organic membrane. Then things went *two* ways. The descendants of LUCA that coupled the acetyl-CoA pathway to the production of methane broke away from their geothermal chimney cloaked in the type of cell membranes we find

2.9 Replication

around archaea. Meanwhile, the descendants of LUCA who coupled the acetyl-CoA pathway to the production of acetate broke away and wrapped themselves in the type of cell membranes that we find around bacteria. Both of these types of membranes are lipid bilayers, but they are of such significantly different composition that it is more plausible to conclude that they had independent origins than to conclude that one type of membrane evolved out of the other. And that's some pretty cool food for thought. The last common ancestor of all currently living creatures on Earth lived glommed onto the surface of a hydrothermal mound or chimney rather than floating about the ancient ocean as a free-living, membrane-bound living cell.

At any rate, now freed from the surfaces of a deep sea hydrothermal mound, those two new forms of life whose basic metabolisms and organic compositions owed themselves to the warm water weathering of silicate rocks were off to conquer the world.

Further Reading

Arndt NT, Nisbet EG (2012) Processes on the young Earth and the habitats of early life. Annu Rev Earth Planet Sci 40:521–549

Kleine T, Palme H, Mezger K, Halliday AN (2005) Hf-W chronometry of lunar metals and the age and early differentiation of the Moon. Science 310:1671–1674

Martin W, Baross J, Kelley D, Russell MJ (2008) Hydrothermal vents and the origin of life. Nature Rev Microbiol 6:805–812

Russell MJ, Nitschke W, Branscomb E (2013) The inevitable journey to being. Phil Trans R Soc B. doi:10.1098/rstb.2012.0254

Sleep NH, Bird DK, Pope EC (2011) Serpentinization and the dawn of life. Phil Trans R Soc B 366: 2857–2869

Sousa FL, Thiergart T, Landan G, Nelson-Sathi S, Pereira IAC, Allen JF, Lane N, Martin WF (2013) Early bioenergetic evolution. Phil Trans R Soc B. doi:10.1098/rstb.2013.0088

Wilde SA, Valley JW, Peck WH, Graham CM (2001) Evidence from detrital zircons for the existence of continental crust and oceans on the Earth 4.4 Gyr ago. Nature 409:175–178

Chapter 3
The Making of Humankind: Silica Lends a Hand (and Maybe a Brain)

Not that you'd necessarily think so if you read the Interwebs or watch a modern Hollywood movie, but human beings are smart. Say what you want about the typical intellect of internet comment makers, but even the greatest of the other living great apes couldn't even dream of being one, much less of being a coder, and what whale or dolphin knows to hope to someday grasp the concept of cinematography, calculus, or the cooking of Italian food. In the midst of even the most chaotic, smog-choked city, with everything it contains and jet planes flying overhead, you can stop and stand bewildered and, hands out, ask: How did we get here? How did we, human beings, get to be smart enough to figure all of this out?

You can also really wonder about those hands, distinctly human, still out in front of you. Where did they come from? You may not hang around enough great apes to know, but our hands are unlike those of all our extant primate cousins. Our fingers are uncurved, short, and braced against the bones of the wrist and have the thumb in famous opposition to them. This means we can grasp with power, pinch with precision, and hold a hammer and use it to really give things a whack.

Manual strength and dexterity at the disposal of a comment-, calculus-, and cuisine-capable brain. In short, we can conceive of and carry out infinity of things beyond the capacity of all nonhuman creatures currently on Earth. Today a typical human being can be taught, not just to push and pull needle through fabric, but the idea of sewing and both that and what it enables you to create. We can learn not just to tap on a keyboard, but that by doing so we could create pain, poetry, prose, music, or an addictive game that could make us millions if enough people want to play it on their phone. Try explaining even a fraction of that to your cat (or orangutan, if you have one).

There is a sense among the archaeologists that the evolution of this manual prowess and this intelligence that make us what we are (and have enabled us to overrun the Earth and massively alter its ecosystems) can be tracked back through the human and the prehuman line by the relics that were left behind. Stone tools in particular. To know the handaxes, choppers, scrapers, knives, ad to not quite infinitum that have been made of amorphous and cryptocrystalline silicate rock during the last several million years and used as extensions of hominin limbs is to know us and how we came to be.

3.1 Stone Tools and Their Makers

To think of stones, if one thinks of them at all, is mainly to have come to mind rounded things smoothed by relentless millennia of wind and water. But when they break, they can be sharp. At some point in the past, somebody who was not quite human figured out that sharp rocks are not just scourges to tender feet. They are useful. And so at first, most likely made accidentally when blunt stones used as hammers cracked, sharp rocks were used for scraping, slicing, digging, and prying things open. Eventually, some hairy Einstein or Einsteina grasped that sharp stones could be created on purpose by bashing one stone upon another. And so began the next three and more than a third million years of punctuated plod toward better ways of breaking stones into tools until we were adept enough to abandon stone for materials more of our own making.

Over that time, what started as smashing two stones and taking whatever sharp edge resulted blossomed into toolmaking industries that required strong, precise hands, relied upon structured, hierarchical plans, took more than five years for an individual to master, and were transmitted faithfully down generations, through and between populations, and from one hominin[1] species to another. At first this transmission happened through simple means like copying. But at some point the toolmaking techniques started to be taught and eventually this teaching included the use of symbolic gestures and sounds that led to fully fledged language.

The fossil record of hominins and their tools implies that the increase in sophistication of the tools made out of silica like flint, chert, obsidian, quartz, and quartzite that fractures conchoidally when struck[2] was tied to evolutionary changes in the architecture of the hands and in the brain in the cranium of hominins. At the same time, hominins' creation and use of stone tools helped to guide these evolutionary changes. The changes to the brain enabling hominins to use tools with their hands and execute the more advanced toolmaking techniques made other things possible as well. Complex language, for example. The kind containing long strings of information inset with clauses and recursion.

---

[1]A hominin is any individual that occurs (or occurred) within the human clade after it diverged from that of Pan (chimpanzees and bonobos) sometime after eight million years ago. This definition currently includes all (extant and extinct) species of the genera *Homo* and *Australopithecus*. It may also include *Ardipithecus kadabba* (in existence roughly 5.5 million years ago), *Ardipithecus ramidus* (around roughly 4.3 million years ago), *Orrorin tugenensis* (extant 6 million years ago), and *Sahelanthropus tchadensis* (which existed 7–8 million years ago, close in time to the human-Pan split).

[2]Conchoidal fractures are the sorts of half clamshell-shaped breaks you might have seen on the surface of a rock like obsidian or chert.

3.1 Stone Tools and Their Makers

In these ways, we have silica to thank for us not being today just another one of several isolated and narrowly distributed primate species snoozing blissfully on the savannah and occasionally feeding the lions. Curse or congratulate silica as you see fit.

3.1.1 The Earliest Stone Tools

The earliest evidence we have for the use of stone tools by hominins is 3.4 million years old. Its form? Scratched bones, namely the fragment of a rib of a cow-sized ungulate and the fragment of a femur of a goat-sized bovid[3] found in a bed of sand at Dikika, Ethiopia. The small number of faint scores visible to the naked eye and the greater number of microstriations upon these bone fragments are consistent with cutting and scraping done to denude the bones of flesh. That required the use of a hard, sharp-edged tool. The bones have also been smashed, perhaps by a blunt stone, as if to release the marrow within. In either case, it also indicates the intentional use of a tool.

This claim, made in 2010, set paleoanthropologists aflame, many with scorn. That such ancient, so very ape-like hominins as lived 3.4 million years ago used stone tools was *Outrageous! Impossible!* and other splutters. For thousands upon thousands of stone tools had been found over the years, but none anywhere near that age and none with the bones at Dikika. If some hominin from those days had been using stone tools, archaeologists would have surely unearthed some where at some point. Something other than tool use must have been to blame for the scratches.

But not everyone was unconvinced.

And so the state of the debate remained until the spring of 2015. That April, a group of anthropologists announced that they had taken a wrong turn near the western shore of Lake Turkana in Kenya and stumbled on the remnants of a 3.3 million year old toolmaking workshop (or, more realistically, a briefly occupied campsite where some toolmaking had occurred). Within only two or three field seasons, this site proceeded to yield roughly 150 specimens of purposefully made flakes of silicate rock which would have been intended for use as tools, the silicate rock cores they were struck from, and the anvils that were used during that process.

These Lomekwian tools, such as the one illustrated in Fig. 3.1, were an earth-shattering find. They are hundreds of thousands of years older than the previously oldest stone tools found and they absolutely and unquestionably could not have been made by a human being (if your definition of human restricts itself to members of the genus *Homo*, which did not come about until about half a million

---

[3]Ungulates are mammals with hooves. Bovids are a family within the ungulate clade and include cattle, goat, sheep, bison, and wildebeests but not horses, rhinoceroses, camels, giraffes, or hippos, to name a few.

Fig. 3.1 Sketch of an idealized Lomekwian flake (intended for use as a tool) broken off of a core by striking the core directly against an anvil

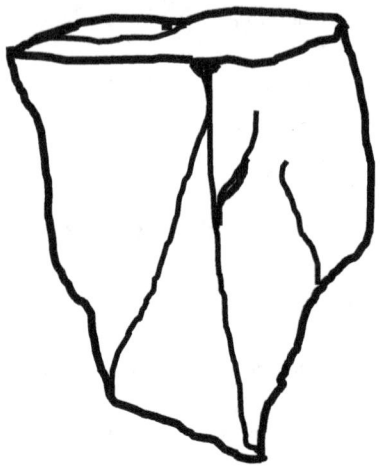

years later). Finally, these Lomekwian stone tools are of a different style than all previously known stone tools. Simpler than the tools of younger toolmaking industries more familiar to us, they grant us a view of a link in the chain closer to that first sharp stone picked up and used by some prehuman hand.

Some of these Lomekwian tools were made by hitting a rock (the tool-to-be) against an anvil (also of rock) held steady on the ground or in the lap. Others were made by hitting the tool-to-be with a hammerstone while the tool-to-be rested upon an anvil. It is easy to imagine that both of these methods arose from the pounding of plants, nuts, or bone to prepare them for eating, an activity that wouldn't intellectually tax a chimpanzee. (We know, because they do this kind of stuff all the time.) How long it took to get from the first stone picked up and used as a hammer to the first stone that was purposefully struck upon another to create a sharp edge, however, remains anybody's guess.

Who would have made and used these oldest known indisputable stone tools? Given the times and places, the perpetrators of the scratches on the old bones at Dikika (if you accept that the scratches were perpetrated) and the makers of the tools at Lomekwi could have been *Australopithecus afarensis* or *Kenyanthropus platyops*.[4]

Lucy of ancient fossil skeleton fame was a member of *A. afarensis*, although she lived 200,000 years after the feast at Dikika was eaten and 100,000 years after the tools at Lomekwi were made.[5] Large bodied, small-brained, bipedal, and routinely reconstructed by museum curators to have the bright eyes and big grins of fairly

---

[4] Or one of the other species existing at that time that we don't yet know about.

[5] *A. afarensis* has been found in the fossil record so far from roughly 3.9 million years ago to 3.0 million years ago.

happy chappies, *A. afarensis* is one of several Pliocene[6] candidates to have given rise to the genus *Homo* of which several million years later, by likely way of *Homo habilis*, *Homo ergaster* (sometimes known as *Homo erectus*), and *Homo heidelbergensis*, *Homo sapiens* is a member.

The other known possibility, *K. platyops* is a controversial species, the sole one in a contested genus. What we have of this species is a 3.5 million year old skull, distorted by its long residence in the ground, found and described at the end of the twentieth century by the team of Maeve Leaky, née Epps. Little more can be said about *K. platyops* except that the hunt for additional specimens continues (and was what had brought the researchers to Lake Turkana in the first place). Even just a few more pieces from *K. platyops* would help to determine if the species truly deserves its own genus or if it ought to be placed in *Australopithecus*. Actually, that's aiming for the moon. More bits are needed just to settle the basic question of *K. playops*. Was it a thing at all or is the fossil just of a weird looking member of *A. afarensis*?

3.1.2 The Oldowan Industry and Its Practitioners

The archaeological record being what it is, we must jump forward 700,000 years in time from the Lomekwi find for the previously oldest known evidence of stone tool use: more stone tools. The chipped stones found at Gona, Ethiopia, are the oldest known artifacts of the Oldowan Industry, a tradition of tools and of toolmaking named for the examples that have been found in the Olduvai Gorge. Oldowan style tools, such as the chopper sketched in Fig. 3.2, were produced from 2.6 million years ago until 1.4 million years ago. That's a span of 1.2 million years, what we modern humans consider an incomprehensibly long time for a fashion not to have changed or for a set of designs not to have been notably improved.

The first practitioner of the Oldowan Industry was probably *Australopithecus garhi*. Like *A. afarensis* and *K. platyops*, *A. garhi* was not a member of the genus *Homo*. So the discovery of its remains in association with Oldowan tools was, years before the discovery of the older Lomekwian tools, a bombshell. Making and using stone tools was supposed to be a key marker for inclusion in the genus *Homo*, the thought being that by the time a species was smart enough to make stone tools, it was human, even if it wasn't as far along in its humanness as *Homo sapiens*. But, skeletally speaking, *A. garhi* was not human enough to be humanity (again, if your definition of human is inclusion in *Homo*). Further, owing to some retrograde features of its face (like big teeth), it is even unlikely to be a direct ancestor of ours,

---

[6]The Pliocene is the period of time that ran from 5.3 to 2.6 million years ago. As with other Epochs of the Cenozoic Era that began 65 million years ago, it represents a downward step in the progressive cooling of climate from the greenhouse warmth and lack of continental ice of the Mesozoic to the cyclic glaciations of the Pleistocene (2.6 million to 12,000 years ago) and the interglacial (but still notably glaciated) conditions of the Holocene (12,000 years ago through today).

Fig. 3.2 Sketch of an idealized Oldowan chopper

as that would require big teeth in one species becoming small teeth in the next becoming big teeth in *A. garhi* then going back to the small teeth when becoming *Homo*. Instead, *A. garhi* probably represents one of the many offshoots off the line that was leading to us.

Which brings us to an important point. Not long ago, the state of our understanding of the human family tree would have appealed to a minimalist. Not too many different species of hominins had been discovered. While the line of descent to ourselves that we had sketched out was not complete, it also wasn't clotted up with jumbles of aunts and cousins to our species, nor with those species' aunts and cousins. These days, though, the twists and turns of the hominin family tree grow leafier each year. The more we dig in the dirt, the more it becomes clear that more hominin species roamed through the Pliocene and Pleistocene than we previously realized and many more species may remain to be found. It is an intriguing, if somewhat is-it-something-that-we-did sort of thought, that the lonely position of being far and away the most intelligent species on the planet is a recent and unprecedented turn of events. Eventually, with enough hard work and good luck, we will have in hand enough fossils for us to say which hominin species is distinctly which, how many hominin species there were, who led to whom, who was just a dead end, and why there were so many different kinds. (And, maybe, how we came to be the only one left.)

But today we're stuck still arguing about basics, such as which species served as the transition from the *Australopithecus* genus (with its cranial capacity of 330–350 cubic centimeters, legs and arms of relatively similar length, large faces and jaws, and bipedality but with curved toes) to our genus *Homo* (with its cranial capacity larger than 510 cubic centimeters, notably longer legs than arms, small faces and jaws, and bipedality with modern feet) roughly 2.8 million years ago. There are several possibilities still in the running, but it is entirely plausible that the correct candidate remains undiscovered in the ground.

After *A. garhi*, the next known maker of Oldowan tools was *Homo habilis*, a species that existed from 2.4 to 1.4 million years ago. That means there were at least two different species who made the same type of tools using essentially the same techniques. This is pause for thought. How could two species so different from each other that one was human and the other was not, have had in this sense the same culture?

There are a couple of possibilities. *H. habilis* could have inherited the Oldowan toolmaking tradition directly from a toolmaking ancestor (although not *A. garhi*, because, as we said, this species has been ruled out as an ancestor of *Homo*). This would only require that the toolmaking behaviors and knowledge were handed down from generation to generation. Or maybe *H. habilis* picked up the Oldowan habit via cultural diffusion. Some members of *H. habilis* might have, for instance, figured out how to make Oldowan tools by watching them get made by members of a species who already knew how to do it. Or they could have been actively taught by individuals from the other species via nonverbal and non-gestural means. If you want to get really wild (and be at great risk of being wrong), you could postulate the involvement of finger pointing and interjections in the lessons. Or perhaps a bit of industrial espionage occurred in the Pleistocene, with *H. habilis* reverse engineering the toolmaking techniques from tools happened upon or stolen from a rival species.

How sophisticated were these Oldowan stone tools and how smart would you have to be to make something like the stone chopper shown in Fig. 3.2? Modern day chimpanzees can knap stones accidentally when they use them to crack nuts and some bonobos in captivity have been taught to create a sharp tool by smashing one rock on another, so at least as smart as chimps and bonobos. But, actually, smarter, for Oldowan stone tools are things significantly beyond what chimps and bonobos can make (as also were the Lomekwian tools, though by less of a stretch). The Oldowan tools include large scrapers (with a sharp face used for working wood or hides), choppers (that had an unworked end that fit in the hand and a sharp edge on the other for cutting, hacking, scraping, and chopping), and smaller tools such as awls (used for puncturing). Minimally shaped and clunky, these tools can look only half emerged from their rounded rock parent. Nonetheless, Oldowan stone tools were intentionally made and to at least some degree planned. Often originally river cobbles, the larger of these tools were made by direct strikes of a spherical hammerstone onto a second stone of basalt, obsidian, quartz, or quartzite, to chip material off the sharp end until it attained a reasonably useful shape.

3.1.3 The Acheulean Industry and Its Practitioners

Although the Oldowan tools and the techniques for their production improved somewhat over the 1.2 million year lifetime of the industry, about 1.8 million years ago a sharp and sudden increase in the beauty and ergonomics of the tools occurred. *Homo erectus* had arrived on the scene and refined the Oldowan Industry into the more sophisticated Acheulean Industry.

Whereas to look at an Oldowan tool, such as the handheld axe of Fig. 3.3, is often to feel that someone broke one end of a rounded stone and got a sharp edge, to look at an Acheulean tool is to be struck by the feeling that the human who made it knew exactly what she or he wanted to make and had the dexterity, skill,

Fig. 3.3 Sketch of an idealized Acheulean handaxe

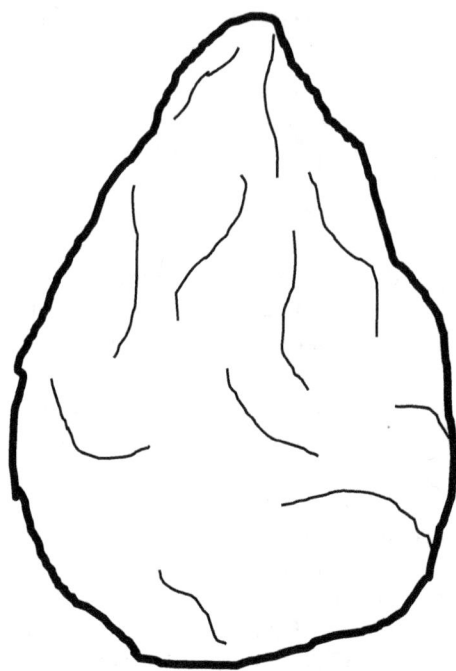

knowledge, and experience to pull it off. The humans who made Acheulean tools must have had a strong sense of aesthetics, that firm appreciations of what makes an object pleasant to use, hold, and behold. These Acheulean tools were often bifacial, that is, worked on both front and back sides. This was new. The Acheulean tools were also satisfyingly symmetrical and the end that was held in the hand was also now extensively shaped (presumably because someone had by then figured out that you could do more than just chip out sharp edges). The holding end was no longer restricted to being as thick and unwieldy as the cobblestone the tool had been made from.

Prototypical Acheulean handaxes, for example, are relatively thin and when faced flat side on, have the unmistakable shape of a teardrop. (They could almost be pendants, such is balanced regularity of their shape.) They were held in the hand (hence their name) and used for a multiplicity of things from butchering and skinning to digging. In what is perhaps also the first clear case of humans having standards, the toolmakers did not make Acheulean handaxes from obsidian, a material available but too brittle to hold up to the demands of handaxehood. They used more durable cryptocrystalline rocks exclusively.

Another key feature of the Acheulean Industry was the production of intermediate shards, known as flakes, that could, with refinement, be turned into tools. Instead of whacking a tool directly out of a rock, you could chip off a large piece of that rock and then shape the piece (already somewhat tool-shaped) into a tool. Although some of this had gone on in the Oldowan, if only opportunistically (one imagines a large flake flying off during knapping and landing on the ground and then a *H. habilis* hand coming up to scratch a *H. habilis* head as it occurs that with a little trim here and a little trim there, that flake could be made useful). By the Early Acheulean, the creation of flakes that were meant to be made into tools was so common and so clearly intentional that it screams a seismic shift in perspective and an increase in the complexity of thought.

Before, the sequence had been simple. The toolmakers wanted to shape a stone into tool. Now the toolmakers wanted to cause large flakes to crack off the original stone so that they could shape these large flakes into tools. Those flakes are like subordinate clauses with subordinate clauses:

I will hit the stone [to yield a flake (that can be made into a tool)].

Was this the first identifiable instance of the recursive thinking that is reflected in the glorious grammar, syntax, and complex-compound sentences of human languages? Some say so.

In the Late Acheulean of 700,000–250,000 years ago and to which not only *H. erectus* but also our immediate predecessor *Homo heidelbergensis* contributed, the refining flaking that was the final step in the production of a tool was itself refined by the innovation of soft hammers. This use of antler or wood for hammering meant that these final hits on the tool being created could be precisely controlled. Broader, thinner tools resulted because it was now possible to remove flakes that ran more than halfway across the flat surface of the nearly finished tool without removing too much material from the edge. Now no one had any excuse for making a chunky, clunky stone tool ever again.

3.1.4 Neanderthals and the Levallois Technique

Starting 300,000 years ago, Neanderthals got into the stone toolmaking game.

Ah, *Homo neanderthalensis*. They have gotten such bad press as dirty, grunting cave men of exactly the sort you've had to suffer through dating once or twice, but in fact they were quite a bit like us. For instance, they too evolved directly from *H. heidelbergensis*. Although, they appeared about a hundred thousand years earlier than we did, our most recent shared ancestor, as calculated from our genetic difference, existed much earlier, about 600,000 years ago. This suggests that *H. heidelbergensis* had split into at least two separate populations. One population went on to give rise to Neanderthals up in Eurasia, while the other went on to spawn *Homo sapiens* down in Africa. Species-wise, that makes us more like siblings than cousins.

Neanderthals' stone toolmaking specialty was the Levallois technique of what has been termed the Mousterian Industry. With the Levallois technique, which requires many years of learning and practice to master, Neanderthals shaped an initial stone using a pestle-shaped hammer to knock off material. When ready, this prepared core was domed on the upper side but flat underneath, reminiscent of a turtle shell. The toolmaker then created a striking platform, also through

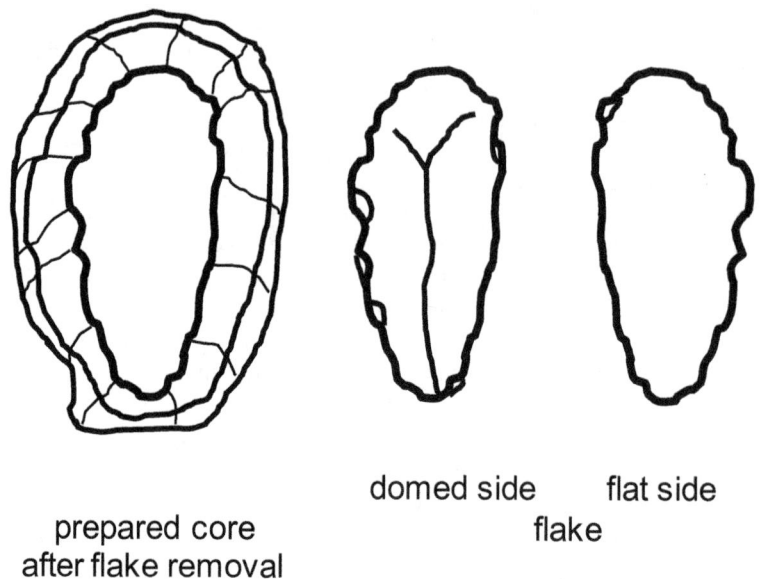

Fig. 3.4 Sketch of an idealized Levallois core and flake (a denticulate scraper) detached from it

hammering, such that when it was struck, a flake of prescribed size and shape broke off of the top of the core, as shown in Fig. 3.4. This flake was generally plano-convex (like the core, flat on the bottom and domed on the top) and, because of the faceted topography of the core, jagged. If the toolmaker had shaped the core properly, further refinement of this "intended flake" would not be necessary before it was used as a tool. The stone tools made this way included handaxes, scrapers, knives, and notched flakes (also known as denticulated flakes) that were possibly used as saws.

If you were to watch someone making stone tools with the Levallois technique, it might seem simultaneously efficient and wasteful. On the one hand, it's fast. It might take five minutes to go from an unworked rock to the first finished tool. It might not be beautiful or symmetrical like an Acheulean tool, but it will be sharp, serrated, strong, durable, and highly useful. And if work on the core continued, three to four more finished tools would follow in short order. On the other hand, reams and reams of shards of rock would have shed off the core as it was shaped, creating quite a lot of pleasantly tinkling glassy rock waste.

Despite this seeming material flagrancy, however, the Levallois technique was a better use of resources than earlier means of making stone tools. Doubtless otherwise it would have been abandoned. The nodules and cobbles of chert and other cryptocrystalline rocks that were large enough to serve as cores would have been a limited resource, potentially requiring a walk of days to acquire.

Pshaw, you say. The Earth is full of rocks. But if asked to supply a half to full football-sized nodule of chert, would you know where to get one? Even in a world

3.1 Stone Tools and Their Makers 39

not covered in farms, concrete, and buildings, usable rocks would have outcropped at only a minor number of locations within the range of any given group of Neanderthals.

The key to the efficiency of the Levallois technique, in terms of its conservative use of material, was that a single core yielded multiple tools. What remained of a prepared core after the detachment of the first intended flake could be briefly touched up and used to create a second flake, and then a third, and so on until the core was too small for the detachment of any more useful flakes. Thus the invention of the Levallois technique doubled the efficiency of conversion of raw stone into tool compared to Acheulean methods. Some of the larger shards sloughed off during the process of shaping a Levallois core (unintended flakes perhaps you could call them) could also be retrieved and knapped into service as smaller tools, further cutting down on waste.

Neanderthals also had the bright idea to combine their stone tools with additional parts. They made some deceptively cute, round, stone balls, for example, that they may have paired with rope and hurled to bring down game, much the way modern day bolas are used. By 200,000 years ago, Neanderthals had also figured out how to make pitch, an adhesive made by baking, under oxygen-limited conditions, birch bark, or conifer resin with finely ground ochre (minerals, like hematite, goethite, and limonite, that are iron oxides or iron hydroxides). One way to do this would have been in a fire in a sealed vessel. Neanderthals used the resulting slowly setting goo for a number of things including, never mind the boat making, hafting[7] stone points onto the tops of wooden shafts. This resulted in spears that were for the first time in prehistory more than just sharpened sticks. These spears were also the first known foray of humankind into uniting multiple disparate elements (a point, a joint, and a handle) into what must thus be called a compound tool. They predate by tens of thousands of years *H. sapiens'* independently originating first grasps at the same.

3.1.5 Homo sapiens

Also about 200,000 years ago, *H. sapiens* appeared in Africa. As had been the case with the Neanderthals, when *H. sapiens* first broke off from *H. heidelbergensis*, they made Acheulean style tools. And as with the Neanderthals, this gave way to the production of stone tools using the Levallois technique. Over time, though, a multiplicity of further toolmaking techniques and traditions developed throughout the ever expanding geographic range inhabited by *H. sapiens* in ways that simply did not happened with Neanderthals. Finally, 50,000 years ago, the pace of cultural and technological development by *H. sapiens* exploded, as if having finally exited some millions of years long lag phase to arrive at the start of the steep part of an exponential curve we are still on.

---

[7]Strongly fixing.

Fig. 3.5 Sketch of an arrow made with multiple microliths

Why this sudden, thrilling rate of advance and why hadn't the Neanderthals managed it despite their hundred thousand year head start? Maybe one last critical bit of brain evolution had occurred in *H. sapiens* that made us, 50,000 years ago, suddenly so very much smarter than the Neanderthals, whose isolated populations were by that time declining and inbreeding themselves out of existence. Or maybe we weren't so much smarter than they were, just luckier in tropical and subtropical Africa than they were in a Europe more often than not half covered by a sheet of ice kilometers high and thousands of miles in length. Perhaps under the clement African conditions, we had the opportunity to become so numerous that by 50,000 years ago the population size and density of *H. sapiens* crossed the threshold necessary for rapid cultural and technological development. Things got so crowded everybody started to bump into everybody else, encouraging the exchange of ideas and innovations, at least between the survivors, and then the spread of the further ideas and innovations that sprung from them.

By this magical moment 50,000 years ago, *H. sapiens* was deeply engaged in making and using hafted composite tools with replaceable components, such as the arrow shown in Fig. 3.5. Now the main focus of stone toolmaking was on little parts that could be combined with other parts and an adhesive to create a tool. This required the production of microliths, bladelets and other small points that could be fastened onto shafts to result in spears, arrows, harpoons, and other devastating weapons and similarly complex tools.

We also began to invent and work with new materials. Fibrous textiles, for starters. The first known sewing needles date to 40,000 years ago, the oldest known dyed flax fibers (which are used to make linen) to 36,000 years ago, and the oldest known clear evidence of woven cloth to 27,000 years ago. And we got into art. The earliest cave paintings date to 40,000 years ago (although some of these older ones could have been done by Neanderthals), animal and exaggerated female figurines to 40,000 to 35,000 years ago, and such figurines molded in clay and fired to produce the oldest known ceramic objects to 30,000–25,000 years ago. In terms of cognitive capacity and behavior, humanity was now officially modern (although there is a habit of calling them early modern humans instead of fully modern humans).

3.2 Hands and Brains

Put into chronological perspective like that, the stone tools of the last 3.3 million years are a record not only of themselves but of what the hominins that preceded us could do with the hands and the contents of the skulls that we sometimes find preserved alongside them. What can we add to this with the hand bones and skulls themselves?

When we have a complete enough set of specimens, hand bones and the marks left by the musculature that was attached to them, yield a direct view of manual dexterity down through the ages. But fossil craniums, distorted or not, don't reveal much more about the brain beyond how big it was. The increasing artfulness of the stone tools and the techniques used to create them (which we can figure out from the tools themselves by figuring out how to make them) remains one of our main insights into hominin intellect over time.

Some of the toolmaking methods, and we're thinking in particular of the Levallois technique, stop you cold and make you think: Wow, I would have never thought of that, not in a million years.

3.2.1 Give Us a Hand

If you want to make a stone tool, and especially if you want to make one skillfully, you need to be able to hold tightly to both the stone you are hitting and the stone you are hitting it with. You also need to deliver a forceful blow. The modern human hand is as if it were designed for this, with the shape, position, and musculature of the thumb and fingers a good ways toward optimized for squeeze.

If you go to West Africa (or to a video on the internet), you can watch chimpanzees smashing large rocks down on nuts they want to eat. It is incredible. They can barely hang on to the rock. Their long fingers are ludicrous and their thumb doesn't seem to be able to act much in opposition to them. The whole thing seems so ineffective.

It's much the same with the video of Kanzi, the bonobo who has been taught by humans, making a sharp stone. He needs one because he must cut through the taut lash that holds the lid of his food box closed. (Mean, manipulative anthropologists! It's clearly a drill he's familiar with.) Kanzi sits down and starts smacking two pieces of chert together. But where are the power and the precision? This is like watching a baby tapping blocks. The brain, the hands; neither seems to have much of a grip on the process.

After several impacts, a shard breaks off one of the rocks. Kanzi drops the rocks, picks the shard up, waddles over to the food box, and begins to saw on the line. Meanwhile, you conclude that even if the nut-crushing chimps and Kanzi the "toolmaking" bonobo had the brains to do better, they couldn't because their hands aren't up to it. They can all only just barely hold the hammerstones and Kanzi can barely use his saw. It's enough to make you wonder where all these hands came from. Human, chimpanzee, and bonobo. So similar and yet so different in their capabilities.

In a fairly literal sense, the hands all came from the last common ancestor of all the great apes that are currently living (these great apes being *Homo sapiens*, chimpanzees, bonobos, the eastern gorilla, the western gorilla, the Bornean orangutan, and the Sumatran orangutan). Based on genetic differences among these living species of great ape, this last common great ape ancestor lived about 18 million years ago. The branch of the family tree that led to orangutans was the first to split off of the line leading to ours, followed by the branch leading to gorillas, and then finally the branch leading to chimpanzees and bonobos. Bearing this in mind, the features of the hands of modern great apes can be sorted into those that occur across multiple branches (and are therefore likely ancestral) and those which are more uniquely possessed (and therefore more recently derived). Based on this, the hand of that last common ancestor was probably most similar to those of the modern day chimpanzees, bonobos, and gorillas.

Among nearly 20 noteworthy features whose descriptions mainly only an anatomist could grasp, that hand of 18 million years ago likely had fingers that were long relative to the thumb. This is something we've already noted is not true of our hands. It also had curved proximal phalanges (the set of four finger bones connected to the solid, central part of the hand). Ours are straight. The distal phalanges of that 18 million year old hand had narrow apical tufts (that is to say, that hand had weirdly thin and pointy bones at the tops of the fingers). And, unlike our hands, that hand had no protrusion, called a styloid process, at the base of the third metacarpal (the bone that sits below the middle finger) to brace the hand against the bones of the wrist.

How did our hands diverge so much from this starting point, especially when the hands of gorillas, chimpanzees, and bonobos didn't really?

The agent was natural selection. To maintain within a population a set of features, such as the exact hand of our 18 million year old ancestor, requires selection pressure or pressures for those features. Maybe they make you great at gathering food or at swinging through the treetops. Because you have those hands that serve you so well, you produce more offspring that themselves survive to produce offspring than, say, your cousins whose changed hands result in frequent 50-foot falls or loss of half of the food they gathered. Their changed hands, a serious handicap with sometimes fatal consequences, will remain uncommon in the population.

But selection pressure is a two-sided coin. Remove or lessen the pressure and the features it was maintaining genetically drift away.[8] If traditional hands and changed hands are roughly equally useful for keeping yourself alive and well fed enough to raise successful offspring, natural selection won't weed either out of the population.

The third option is that the modified hands turn out to be more useful than the original ones. A little shift in the thickness or attachment of a muscle, a small alteration to the shape or position of a bone that perhaps results in the greater efficiency or success of actions, like food gathering, that were already being done. Or maybe the modification makes it possible to do things differently or to do things that haven't been done before. As long as the end result is greater success in the

---

[8]Hence the amazingly large number of people who need glasses even early on in life. Perfect vision has not been critical to survival and reproduction of humans since we started living in cooperative groups or maybe since we started plowing. Take your guess.

3.2 Hands and Brains

production of offspring that successfully reproduce (and so on), those modifications may become common within a population. Depending on exactly how much greater the success (relative to the success of those with the unmodified hands) and the size of the population, it could take decades, centuries, or thousands of years, but eventually everyone will have the modified hand.

Of course the evolutionary process is somewhat chicken and egg, occurring as a little bit of this encouraging a little bit of that which then feeds back and results in more this, etc, and who knows which kicked the whole thing off. But we may still ask what selection pressures lost, lacking, or gained resulted in the human hand. Our hominin ancestors were disinclined to knuckle walk, for one, so our hands were under no evolutionary pressure to be good for that. This true bipedality freed up a lot of ground in terms of the possible architecture of the hand. Ditto for our having stopped living in trees. Our fingers could get short and our thumbs could become more opposable to them because having a hand that was bad for tree-swinging was no longer a problem. Lastly, our line of hominins has long had a penchant for picking things up and manipulating them (or hitting things with them). If this had value in terms of survival and reproduction, then grip strength and nimble fingers would have been selected for.

Enter tools made from silica. If they made it easier for hominids and hominins to feed themselves, protect themselves, or solve other problems of life in the Miocene, Pliocene, and/or Pleistocene, any genetic remodeling of the hands that favored making and using stone tools had a good chance of being retained and then increasing in frequency within a population until the genetic coding for hands less capable of making and using stone tools died out.

When exactly all of this happened and how many steps were involved remains a jigsaw puzzle missing most of its pieces. We have found virtually no fossil hand bones older than 3.2 million years, or at least not any sets complete enough to support sound conclusions. To make matters worse, 3.2 million years old is too young. The first big changes that had to occur as our distant ancestors picked up stone for the first time and started to use them as tools and then to fashion them into tools had to have occurred earlier than this. For by a bit before 3.2 million years ago, as the finds at Dikika and Lomekwi testify, hominins were enthusiastically engaged in both stone tool use and production.

The handful of hand bones of *Australopithecus afarensis* we have from that time tells us that the fingers of the hominin hand had already shortened somewhat relative to the thumb and some of the joints of the wrist on the index finger side of the hand were also sort of human looking. The *A. afarensis* hand would have had much better grip strength and toolmaking dexterity than the 18 million year old hand. This *A. afarensis* hand was, just by brute force logic, at least the least derived hand necessary to create and use Lomekwian and Oldowan stone tools. Whether or not this hand was already several changes beyond that, however, remains unknown.

The next big modification of the hominin hand (changes to the fingertip bones, to the thickness of the thumb, and to the joint that connects the base of the thumb to the wrist and is utterly critical to the functioning of the thumb) appeared in early specimens of the genus *Homo* between 2.0 and 1.5 million years ago. Now hominins could really grip both a rock and the hammerstone they were hitting it with. Although the sparseness of the fossil record means that these features could have originated somewhat earlier, it is interesting to note that the sizable leap from the Oldowan to the Acheulean toolmaking tradition occurred 1.8 million years ago. In other words, at roughly the same time. Considering that the Acheulean Industry, with its thin, finely wrought, and beautifully symmetric tools, is light years ahead of the Oldowan, those changes to the thumb, in concert with whatever changes happened between the brain of prehuman hominins and that of human hominins, was likely revolutionary for toolmaking.

The final major change in the hominin hand, involving changes to the more pinkywards side of the wrist, occurred relatively recently, no later than 800,000 years ago. This means it was common to Neanderthals and *Homo sapiens*, both of whom brought stone toolmaking techniques out of the Acheulean, first by inventing things like the Levallois technique of the Mousterian Industry and then a multiplicity of techniques beyond it. But it is not clear that the Mousterian Industry needed stronger, more agile hands than the Acheulean Industry had required. This leap forward may have been more one of intellect—being smart enough to have the insight to invent the Levallois technique which seems sloppy but is a super quick way to make tools that are as effective or moreso than the beautiful, labor-intensive tools of the Acheulean. Perhaps this particular set of significant modifications to the human hand were related to the use of the tools after they were made.

Whatever the case, the grip, pinch, and nimbleness of the hands you have in front of you are a good (although not perfect) indication of those of human hands of several hundred thousand years ago. Playing baseball, the piano, and video games; sewing, knitting, weaving, and crocheting; and making all of the stone tools we know from hominin prehistory would have been *physically* possible for Neanderthals and the earliest individuals of *H. sapiens* thanks to millions of years of selection pressure for hands that could knap flint and handle the resulting stone tools.

3.2.2 If I Only Had a Brain

If you're anything like us, at this point, you totally want to make a stone tool. And if you're anything like us, you have nothing in the way of previous experience. Or, well, maybe you've seen some arrowheads in a basket on the counter of a national park gift shop in the American west. Or maybe you've seen some ancient handaxes in a museum of antiquities in Europe. At least thus having a vague idea of something to aim for, you could try to figure out how to make one by that process with that terribly technical sounding name of reverse engineering. You'd learn faster, however, and would do a better job of reproducing the tool you were aiming for if you watched someone make one and then aped their physical movements (imitation) and their handling of the objects they're working with (emulation). But if someone guided your attempts to make a stone tool, perhaps by moving your hands or rotating the stone into a good position before you hit it, you'd learn better and faster still. And if the teaching occurred with symbolic means of communication, such as gestures (like pointing) and sounds (anything from grunts, to interjections, to sentences), you'd pick up the technique the fastest and most accurately because someone could at least to some extent convey what you needed to be doing and why, where you were going right, and where you were going wrong. Thus that person could share with you tricks and tips accumulated over years (if not generations) of practice.

We know that to make the progressively more sophisticated stone tools of the archaeologic record required changes to the hands and to the ability of the brain to plan and understand the toolmaking. But the ability of hominins to teach toolmaking to each other likely also played a role in the maintenance of the more sophisticated traditions as well as in the growth of more sophisticated toolmaking industries out of simpler ones.

One study tried to demonstrate this by recruiting a bunch of inexperienced people to make Oldowan tools. The recruits were split into groups and given a few hours to learn the techniques using different methods. Some people were relegated to reverse engineering and others to imitation and emulation. Some were taught using nonverbal and non-gestural methods and the last two groups of people were taught using instructions or gestures. For what its brief duration and use of *H. sapiens* are worth, this study concluded that imitation and emulation were *not* effective means of transmitting the Oldowan Industry. Only teaching was, especially when that teaching included the use of language.

No anthropological study is good without speculation to round out the conclusions. The story here is that perhaps the stasis of the Oldowan Industry over its grand expanse of time and geography means that its practitioners were lousy teachers. Maybe hominins back then had to spend all their time reinventing the same old stone tools because no one could teach them the finer points of making them. Thus these hominins never mastered their toolmaking techniques to the point of being able to improve them or the tools that resulted. And if one of the hominins from this time had managed to innovate, if he or she couldn't effectively communicate, the great idea would have been doomed to die with them.

The Acheulean Industry may have finally arisen when the ability to teach, in at least a rudimentary way, arose in hominins. In humans, in fact, because it was *H. habilis* who brought us the Acheulean. Of course *H. habilis* had to also be smarter and more dexterous than its forbears in order to dream up and execute the Acheulean techniques, but even so, you could be forgiven for thinking, *teaching: older than the world's oldest profession.*

It is fairly easy to imagine steps by which hominins developed the fundamentals necessary for communication. Alterations to the diaphragm, throat, larynx, vocal chords, and tongue acquired for calling and later refined for articulation could easily have been selected for in groups that wandered, hunted, gathered, and otherwise worked together. It's also pretty easy to imagine how the ability to convey information through a growing array of grunts, postures and facial expressions, gestures, words, and, finally, simple sentences (*Me, Tarzan. You, Jane.*) would also have been useful enough to have been selected for. The prodigious expansion in working memory necessary to store and quickly fish up any of the tens of thousands of words? Almost certainly the result of all the other things nomadic hominins had to get good at remembering (like which mushrooms were poisonous and how to get back to camp) as they navigated their way through their territory.

It has been harder to envision how the complicated, highly organized, and still poorly understood neural structures that are critical to producing and understanding language became established and then wired together. This means Wernicke's area, Broca's area, and the entire inferior frontal gyrus of the human brain. Without them, we would indeed be as eloquent as Tarzan.

One hypothesis is that at least some of these neural structures were established through natural selection via the manipulation of stone tools and then co-opted for the speaking and comprehension of complex language. For hands to successfully use a tool (for example, a hammerstone for removing material from a stone during toolmaking), a series of motions (of which, like words, there is a limited set) must be combined in a specific order. This sequence of movements has a hierarchical structure (first do this, then do that) and subsets of sequences embedded within it (for example, repeat this set of movements until that is achieved then go on to the next set).

These strings of movements are analogous to a sentence. In sentences, words are strung together in a specific sequence that conveys specific meaning and often contains recursive elements like subordinate clauses. Thus a neural structure that can create and make sense of such a hierarchical sequence of movements should in principle be able to handle hierarchical sequences of words.

According to the magnetic resonance imaging, Broca's area is the part of the brain of *H. sapiens* that becomes active when someone uses a tool with their hands (or at least imagines using a tool with their hands, MRI machines being too cramped for toolmaking motions). Broca's area is better known for processing grammar that involves hierarchical structures [such as most of the sentences in this chapter (because they tend to be complex [and compound])]. Nonhierarchical grammar (*I am Tarzan*) is dealt with elsewhere in the human brain. This supports the idea that the neural substrates in the hominin brain that processed the structured, hierarchical instructions guiding hands during tool making and use were put to further use processing other structured, hierarchical strings of information. Yes, this means sentences. But it also means music and arithmetic. So you can start thanking silica for Bach, Beethoven, Beyoncé, and Sudoku.

Of course, we're far from sure that the developments in the hominin brain improving its coordination of hands making and using stone tools made it possible for us to express and understand fancy grammar. But it pleases us to think we might thank silica for poets.

Maybe because we worked with silica with our hands, our brains became able to understand that there are many different types of words—nouns, verbs, pronouns, adjectives, adverbs, and so forth—and that they can be strung like beads to convey ideas. And maybe we can blame silica for that first moment when the conjugation of a verb occurred, proof that we hominins had developed a sense of past, present, and future, as well as that outcomes could be conditional. And maybe we can blame silica for complex-compound sentences that reflect an ability to embed subordinate ideas and perspectives into streams of thought. At the point such recursion became possible, hominins could definitely know that they were wondering what someone else was thinking. They could think about what someone else would be seeing from where they were standing and they could consider what it would be like to walk, if not a mile since those probably didn't yet exist, then some distance in someone else's shoes, sandals, or, most likely, bare feet.

Further Reading

Harmand S, Lewis JE, Feibel CS, Lepre CJ, Prat S, Lenoble A, Boes X, Quinn RL, Brenet M, Arroyo A, Taylor N, Clement S, Daver G, Brugal J-P, Leakey L, Mortlock RA, Wright JD, Lokorodi S, Kirwa C, Kent DV, Roche H (2015) 3.3-million-year-old stone tools from Lomekwi 3, West Turkana, Kenya. Nature 521:310–315

Higuchi S, Chaminade T, Imamizu H, Kawato M (2009) Shared neural correlates for language and tool use in Broca's area. NeuroReport 20:1376–1381

Marzke MW (2013) Tool making, hand morphology and fossil hominins. Phil Trans R Soc B 368. doi: 10.1098/rstb.2012.0414

McPherron SP, Alemseged Z,. Marean CW, Wynn JG, Reed D, Geraads D, Bobe R, Bearat HA (2010) Evidence for stone-tool-assisted consumption of animal tissues before 3.39 million years ago at Dikika, Ethiopia. Nature 466:857–860

Morgan TJH, Uomini NT, Rendell LE, Chouinard-Thuly L, Street SE, Lewis HM, Cross CP, Evans C, Kearney R, de la Torre I, Whiten A, Laland KN (2015) Experimental evidence for the co-evolution of hominin tool-making teaching and language. Nature Comm 6: 6029. doi: 10.1038/ncomms7029.

Seemaw S, Renne P, Harris JWK, Feibel CS, Bernor RL, Fesseha N, Mowbray K (1997) 2.5-million-year-old stone tools from Gona, Ethiopia. Nature 385:333–336

Tocheri MW, Orr CM, Jacofsky MC, Marzke MW (2008) The evolutionary history of the hominin hand since the last common ancestor between Pan and Homo. J Anat 12:544–562

Chapter 4
Mystical Crystals of Silica

People split into two different camps—those who believe that crystals have special powers and those who roll their eyes. We (the authors) have long been eye rollers. We are scientists, after all. Acquaintances professing spiritual exuberance for quartz or steeping Himalayan rocks to make crystal energy tea send us into stammers of embarrassment. Crystals are solids surely as mystical as butter. So, alas, the joke was on us when we knuckled down and read up on the scientific behavior of crystalline silica. It's not exactly as the New Agers and several other more traditional traditions have it, but give a quartz crystal a squeeze and it will give off electricity. Who knew? Physical chemists, physicists, mineral physicists, materials scientists, crystallographers, and engineers, for one (or six) and, hey, now we have a lot of modern technology. Let us explain.

4.1 What Is a Crystal?

Before we can explain the powers of crystals, we need to understand what crystals are. So think of a crystal.

What just came to mind?

Those cool, clear, pointy-ended hexagonal prisms of quartz found on the ground and in the display cases of occult and neopagan shops? Or did you think of gemstones that, like diamonds, sapphires, emeralds, amethysts, and topaz, are clearly crystalline? Of course, there are quotidian crystals too, like sugar, salt, and ice. Crystals are as common as dirt. Literally. (Dirt is commonly full of them.) Quartz, feldspars, pyrite, biotite, zeolites, and a multitude of other crystalline minerals lurk there alongside noncrystalline materials like clays, soot, ash, and decaying vegetation.

It's kind of amazing. We're all but swimming in a sea of crystals. But it is also unsurprising; a crystal is simply any solid material whose chemical repeating unit, be it an atom (such as Si) or molecule (such as SiO_2), is arranged in an ordered and repeating structure instead of as a mumble of jumble. How ordered and repeating is ordered and repeating? One of the earliest X-ray crystallographers, and therefore one of the first human beings to have been charmed by the arrangements of atoms in crystals, described the patterns as lace, mosaics, and flowers.

Such strict, lovely order, when it exists in a natural world normally so ragged, seems both improbable and capable of great things. It is wonderful to wonder if such regimentation requires a guiding hand, some all-powerful atomic traffic director with a little whistle and a tiny hat. It is even more inspiring to believe that atoms linked up in three-dimensional geometric arrays possess otherworldly abilities to channel energy.

Well there is indeed a traffic director when crystals form, although not one capable of whistling or wearing hats (nor, for that matter, of having whims or making decisions). Atoms come in difference sizes and charges and this means there are physical limits, attractive forces, and repellent forces at play in the placement of the atoms in a solid. Given enough time, atoms forming a solid will arrange themselves into the structure where all the atoms are in the easiest place for them to stay.

To give you some chemistry for poets, atoms consist of a positively charged nucleus (which contains positively charged protons and neutral neutrons) surrounded by a whirling cloud of negatively charged electrons. The atoms that possess a greater number of protons than electrons are positive. The atoms that have fewer protons than electrons are negative. Roughly speaking, positive atoms repel positive atoms but attract negative ones. Likewise, negative atoms repel negative atoms but attract positive ones. Squishing repellent atoms into close quarters is like trying to force the wrong ends of two magnets together. It does not work very well. When atoms and molecules come together during crystallization, such forces maximize the distance between positive ions and also the distance between negative ions within the structure. Atoms that are attracted to each other will, on the other hand, tend to snuggle up close or even form bonds.

The overall effect of this is that disorder in solids can be more difficult to create or maintain than an order where the forces of repulsion fall somewhere between balanced and minimized. Crystallinity is in effect the structural equivalent of the path of least resistance.

In a boring universe, only one crystal structure would be possible for a given chemical compound. In our universe, the crystal structure of a compound can vary with the conditions around it. Mainly this means pressure and temperature, which, on Earth means it matters where the crystal is. Pressure and temperature at the Earth's surface are much lower than they are deep within the Earth's crust or, below that, in the mantle. A crystal of, for example, the SiO_2 of silica forming near the surface will have a different highly ordered structure than a crystal of silica forming in the ground several miles beneath it.

The different crystal structures of the same chemical compound are known as the compound's polymorphs. The list of the polymorphs of silica includes α-quartz,

4.1 What Is a Crystal?

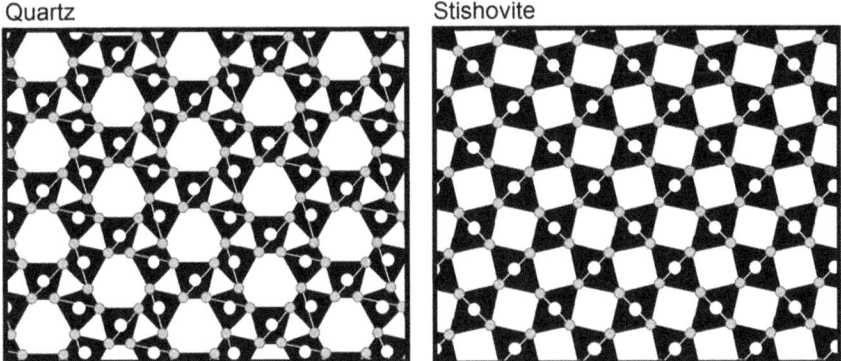

Fig. 4.1 The horizontal cross-section view of the crystal structures of the silica minerals quartz (on the *left*) and stishovite (on the *right*). *White spheres* are silicon atoms and *grey spheres* are oxygen atoms. Note that in the lower pressure (less densely packed) form of silica (quartz) the silicon is coordinated tetrahedrally (i.e., with four oxygen atoms) while in the higher pressure form (stishovite), the silicon is coordinated octahedrally (i.e., with six oxygen atoms), although in both cases the overall chemical formula is SiO_2. This figure is based upon the c-axes of stishovite and quartz presented by Funamori N, Kojima KM, Wakabayashi D, Sato T, Taniguchi T, Nishiyama N, Irifune T, Tomono D, Matsuzaki T, Miyazaki M, Hiraishi M, Koda A, Kadon R (2015) Muonium in Stishovite: Implications for the Possible Existence of Neutral Atomic Hydrogen in the Earth's Deep Mantle. Sci Rep 5:8437 (https://www.ncbi.nlm.nih.gov/pmc/articles/PMC4326963/figure/f1/) and available for use under Creative Commons License Attribution 4.0 International Public License (https://creativecommons.org/licenses/by/4.0/legalcode)

β-quartz, tridymite, cristobalite,[1] moganite, coesite, keatite, stishovite, and seifertite. That's *nine* possible crystal structures for silica and nine entirely different crystals. If you held any two of them in your hands you'd think someone was crazy if they told you they were both SiO_2. The crystals would have different shapes and likely different sizes. One mineral would be notably denser than the other. If you tried to melt or dissolve the crystals, you'd also find they had different melting points and an unequal propensity to dissolve in water.

Unfortunately, we can't reach out and hand you a piece of quartz (that familiar phase of silica formed at geologically modest pressures) and a piece of stishovite (which is formed at higher pressures), but we can show you, via Fig. 4.1, the difference (as viewed two dimensionally) in the arrangement of silicon and oxygen atoms in the two minerals. Stishovite (mainly known to science from specimens found in ground KO'd by a meteorite) has been so compressed, there is no longer room in this silicon dioxide polymorph for silica tetrahedra (pyramidal-shaped units

---

[1]Disappointingly, the mineral cristobalite is named for the place where it was first found (Cerro San Cristobal, Mexico) rather than because it is used in crystal balls. In fact, being opaque instead of transparent, cristobalite is unsuitable for use as a crystal ball. Quartz would be the better choice here.

Fig. 4.2 Stability of the different crystal structures of silica in relation to pressure and temperature

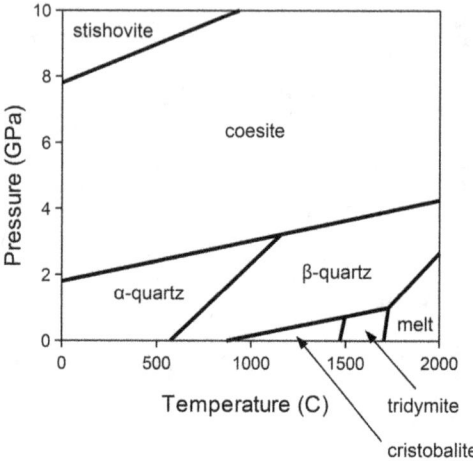

of SiO_4) to exist. Instead stishovite's overall formula of SiO_2 comes about through the interlinkage of octahedra of SiO_6, as can be seen in the right panel of Fig. 4.1. This makes stishovite almost twice as dense as quartz and harder than almost every oxide mineral known to mankind.

Which crystals form under what temperatures and pressures has been studied for a long time. This is useful information if you are a geologist trying to work out the history of the rocks or landscape in front of you. Figure 4.2 maps out what has been worked out for silica. At the "lower" pressures found from the surface of the Earth down to depths in the ground of 50 to 140 kilometers (31 to 87 miles), the crystal that forms is quartz (although it has two slightly different structures, that of α-quartz and that of β-quartz). Because they are the lowest pressure phases, these two types of quartz are relatively roomy in terms of the way the silica tetrahedra are linked together in the crystal (look at all that empty space on the quartz panel of Fig. 4.1, representative of the empty space of other planes within the crystal as well).

When crystallization occurs at somewhat higher pressures, the silica tetrahedra are pressed closer together to form crystals of coesite. As with the even higher pressure polymorph stishovite, you'd never mistake coesite for quartz. It's much heavier, it's not hexagonal, and it tends to be extremely miniature.

You have to go pretty deep into the Earth to get to the pressures that form coesite. Until recently, the only naturally occurring coesite we knew about had actually been formed near the surface, like stishovite, in the shock of a meteorite impact. In the last decade, however, we've figured out that microscopic coesite crystals can be found in those rare pieces of mantle that have ascended from deep within the Earth, raised by rising plumes of magma that also deliver diamonds and zircons to within reach of man. Microscopic coesite can also be found worldwide in crustal rocks that have been subducted to depths deeper than 70 kilometers (43 miles) and metamorphosed under the blistering temperatures and pressures there before being returned to the surface.

4.1 What Is a Crystal?

But still more torture is possible. By the time poor silica is dealing with pressures of 8–10 gigapascals (roughly 80,000–100,000 times the atmospheric pressure at sea level), the silica tetrahedra collapse. The oxygen and silicon atoms of the silica become packed into octahedral arrays (with each silicon being bound to six oxygen atoms instead of the four of the tetrahedron). This is when the crystal that forms is stishovite. Technically, you could say that, lacking silica tetrahedra, this SiO_2 is no longer silica, that stishovite is a nonsilica silicon dioxide.

You may have noticed that the silica phase diagram (Fig. 4.2) is missing the majority of the polymorphs of silica that we listed a few pages earlier. Some of the polymorphs that do not appear are only metastable, meaning that even at their optimal regime of temperature and pressure their crystal structure will slowly rearrange itself into one that is more stable. The other missing polymorphs are simply off the chart. Seifertite, for example, requires pressures in excess of 35 GPa to form (hello really super stupendous meteorite impact) and our chart ends at 10.

Another thing that is not on our phase diagram is something that is also missing from our list of silicon dioxide polymorphs: amorphous silica. This silica is not crystalline.

You know amorphous silica even if you don't know you know it. It pops up in our lives as the semi-precious gemstone opal, as quartz glass (which would be more accurately named silica glass),[2] and as glassy rocks like obsidian (lava that cooled too quickly to crystallize). The word amorphous indicates that there's no order to the arrangement of the silica tetrahedra within the material. But despite this internal chaos, the outer form of amorphous silica is often wondrously and rigorously shaped. For instance, though messily condensed out of a liquid medium and run through with carbohydrates and proteins that serve in part as the mold, amorphous silica produced by diatoms, radiolarians, a lot of marine and freshwater sponges, and some types of soil microbes and land plants is finely wrought, full of precisely placed pores, processes, lattices, spikes, and constellations of crenelations.

Although, like the amorphous silica of diatoms, crystals can grow into the shape of the space they are given, when given no limits, crystals take on an ideal form. The sodium chloride in your salt shaker likes to come as small cubes. A perfect calcite crystal is a rhombohedron. Our friend quartz at its most euhedral is a hexagonal prism with a six-sided pointy top and bottom, as illustrated in Fig. 4.3, while coesite, as we've said, would never be hexagonal.

Can we make some sense of this? The shape, surely, must be telling us something about the underlying crystal structure. Enter the crystallographer, part physicist, part chemist, part mathematician, part spatial visionary genius. Crystallographers shoot X-rays through crystals to see how the beams reflect and

---

[2]However, few of the things we think of as being glass are actually pure silica. What we loosely refer to as glass usually contains, in addition to silica, sodium, calcium, boron, magnesium, aluminum, lead, potassium, and/or other additives that give the material color, alter its transparency, change its sheen, or lower its melting point or otherwise simplify the processing involved in its production.

Fig. 4.3 A quartz crystal showing the three quadrangular prisms of the underlying crystal structure

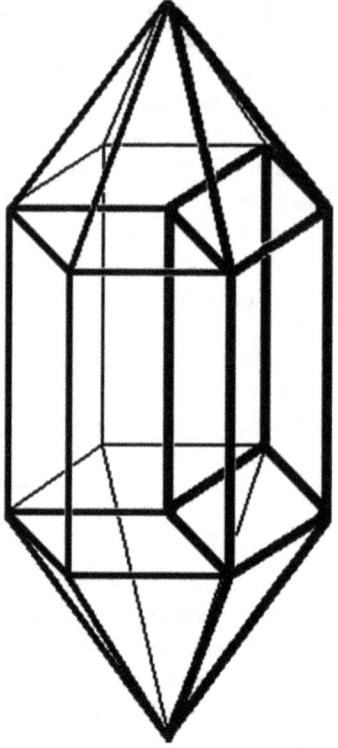

refract their way through the molecules in the material. Coupling the pattern of blurry bars and empty spaces that results with close study of the geometry, symmetry, and angles of crystal faces and of the chemical composition of the crystals, and with mathematics guided by thermodynamics and quantum mechanics, crystallographers discern what cannot be seen by eyes. Crystallographers have fathomed why crystals have their particular shapes and their certain physical, chemical, and sometimes even electric properties.

So sodium chloride crystals come as cubes. Does that mean the lattice of crystalline sodium chloride is a grid of cubes? Sometimes the universe is indeed that dull. The smallest unit you can break a sodium chloride crystal down into and still preserve all of the symmetry of the crystal is a cube containing 13 sodium atoms and 14 chlorine atoms. We would say that this is salt's *unit cell*. When brought together, the edges of these cubes form the gridlines of what is known as the crystal lattice. But please take note, this crystal lattice is conceptual. The atoms inside of the unit cell (as opposed to running along the edge and therefore comprising the conceptual crystal lattice) are not just floating there. They are very much bound into the real, physical structure of the crystal. At any rate, for salt there is a direct translation up from this smallest recognizable crystal unit (i.e., this unit cell) to a salt crystal big enough to see.

4.1 What Is a Crystal?

What about quartz? If an ideal quartz crystal is a hexagonal prism poised between tapering ends, does this mean the unit cell of quartz is a hexagonal prism? This time the answer is no. Quartz can be broken down further than a hexagonal prism and still express all the shape and symmetry you need to create the crystal. The unit cell of quartz is a rhombus (a six-side figure whose every face is a kite) that contains three silicon atoms and six oxygen atoms, as shown in Fig. 4.4. Stack a number of such rhombi together and you end up with a quadrangular prism (a prism whose upper and lower faces are kite-shaped). Join three of these quadrangular prisms together, give them a pointy hat at each end, and, voilà, you have the hexagonal prism of a quartz crystal (Fig. 4.3).

The unit cell, like the lattice, is conceptual as opposed to being a real physical building block. The left panel of Fig. 4.4 just screams this. Note how the superimposed unit cell of the quartz crystal lattice takes no notice of the silica tetrahedra that are the actual physical building blocks of quartz. The unit cell and the crystal lattice which are its edges ride roughshod right through them. This unit cell is

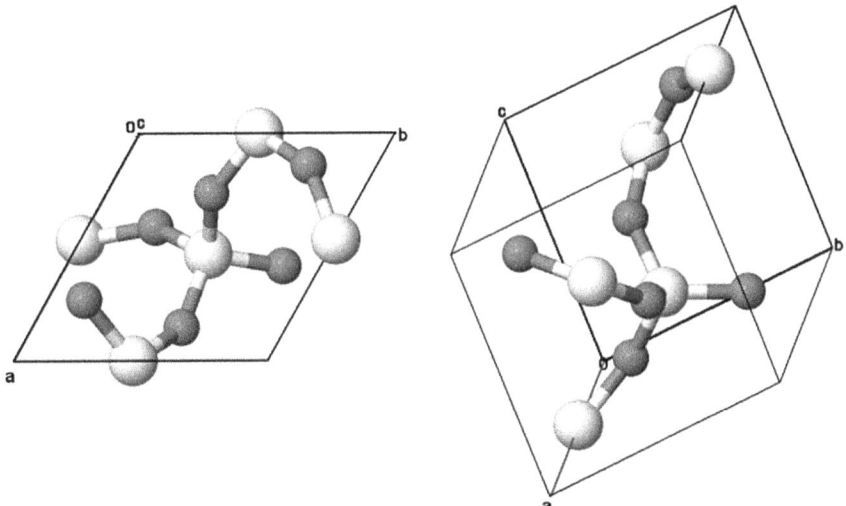

Fig. 4.4 Two views of a unit cell of crystalline quartz. The *left panel* shows top view (looking down upon the unit cell). The *right panel* shows a slight rotation so that the upper pane of the unit cell is the one that was being looked down upon. This gives a better view of the three-dimensional placement of the silicon atoms (the *larger white spheres*) and the oxygen atoms (the *smaller grey spheres*) within the unit cell. Bonds between atoms are shown as *sticks*. The labels *a*, *b*, and *c* mark the different axes of the unit cell (analogous to the x, y, and z you might remember from creating three-dimensional plots in geometry class) of the unit cell as originating from the corner labeled 0. This image has been rendered into *black* and *white* from images obtained from the American Mineralogist Crystal Structure Database (Downs RT, Hall-Wallace M (2003) The American Mineralogist Crystal Structure Database. Am Mineral 88:247–250; http://rruff.geo.arizona.edu/AMS/amcsd.php) based on data for quartz at 1 atm of pressure from Levien L, Prewitt CT, Weidner DJ (1980) Structure and elastic properties of quartz at pressure. Am Mineral 65:920–930

merely a repeating geometric pattern based on the positioning of the atoms within the crystal.

What the shape of the unit cell and the long quadrangular prisms based on them in Fig. 4.3 tell us about quartz, is that even though a hexagon in geometry has symmetry such that its opposite faces are equivalent, the opposite faces of the hexagonal prism that is a quartz crystal are not equivalent. If you number the two outer vertical faces of each quadrangular prism of the quartz crystal in Fig. 4.3 sequentially as 1 or 2, you will find that outer faces that are 1s oppose outer faces that are 2s. This will turn out to be utterly critical for quartz's unexpected electric powers.

That sounds kind of cool, doesn't it: quartz's unexpected electric powers.

Have you experienced them?

That's actually a stupid question. Of course you have.

Unless you've been living in the most remote and isolated area of Amazonia, entirely out of contact with the outside world, you have at some point owned, used, or been subjected to a quartz crystal clock or watch (or any appliance, like a microwave or DVR, that keeps track of the time), a computer, a radio, a telephone (fixed or mobile), a Geiger counter, medical ultrasound, an ultrasonic toothbrush, sonar, and/or an oscilloscope, among many other things. There can have been truly few days of your life when you have not tapped into the piezoelectric power of crystals of quartz. Once you know what it is, it is as hard to imagine living without piezoelectricity as to imagine living without plain old ordinary electricity. Yet we've only known about piezoelectricity for about 135 years and only really mastered it within the last 90.

4.2 Pyroelectricity

The first step on the path toward *Homo sapiens*' first known encounter with the piezoelectric effect was taken toward the close of the fourth century BCE by a fellow named Theophrastus. This student then successor of Aristotle was one of those ancient Greek philosopher-scientists who got to sit around, nose around, and think about *Everything* and boy did he ever. Dust soot over modern science and rational thought and you'll still find his fingerprints. By identifying, describing, and studying as many plants as possible in order to organize them into hierarchically related groups and categories, Theophrastus basically invented taxonomy. He prefigured modern physics when, trying to understand the interrelationships of objects, motion, and time, he viewed time as a necessary side effect of objects being able to change positions relative to one other. He dabbled in the grammar of logic. He even considered animal rights, refusing to eat meat as it was unjust to take life from an animal with the power to sense, reason, and feel. But what we two authors like him most for was his way with rocks.

4.2 Pyroelectricity

For starters, Theophrastus did for them what he did for plants. He identified and attempted to organize a vast number of rocks, minerals, and gems, categorizing them according to their properties, such as hardness, or origin, be it volcanic, organic, or otherwise. As with his work on plants, you can laugh it off as neat-freakishly obsessive. (Seriously, can you imagine the state of this man's closets and drawers?) But organizing and interrelating things on the basis of their properties and behaviors is a powerful approach, one still central to the natural and physical sciences. For once you have things in order, you can start asking questions about the processes leading to phenomena and this is when monumental discoveries are made.

Theophrastus, in this quest to identify and describe rocks and minerals and their properties, subjected rocks and minerals to enhanced interrogation techniques. In particular, he liked to heat them up. Perhaps he was a pyromaniac. More likely it was that back in Classical Antiquity, fire was as cutting edge a tool as there was for probing stones. And Theophrastus did find something so astonishing it defied explanation for the next millennium and half. When you heat certain minerals (in his case, crystals of the beautiful aluminosilicate tourmaline) they become attractive to specks of the nonconductive, for example, sawdust.

How bizarre.

No one knew what to make of it until the mid-1700s when Carl Linnaeus (another obsessive classifier)[3] realized... *electricity*. When tourmaline crystals change temperature they take on a charge. Nowadays we say they become pyroelectric. But this electric charge doesn't last forever. Once tourmaline's temperature stabilizes, the voltage on it dissipates.

Theophrastus didn't know it, and for that matter neither did Carl Linnaeus, but it's all about the surface charges on the tourmaline relative to the distribution of electrons and protons within the crystal. Protons in the nucleus of an atom are what give the nucleus its positive charge and electrons around the nucleus, as you may recall, are negatively charged. Overall in the crystal the positive and negative charges add up to zero, or no net charge. But for some crystals, over very small scales within the crystal there can be imbalance, giving the crystal a polarity (more positive charge on one side, more negative charge on another). In practice, however, the charges that accumulate on the outer surface of the tourmaline crystal neutralize this polarity. When you change the temperature of tourmaline, the distribution of electrons and protons within the crystal shifts enough to produce an imbalance in charge from one face of the crystal to another. That is, at least until the compensating surface charges adjust themselves and cancel the polarity out. Hence the tourmaline crystals only generate electricity when they are heating up or cooling down.

---

[3]Carl Linneaus, as in the father of modern taxonomy, a fellow who no doubt knew his Theophrastus inside and out.

4.3 Piezoelectricity

But we're getting ahead of the story of quartz's unexpected electric powers. You now know more than did the brothers Curie (Jacques and Pierre) when they were first working as laboratory assistants at the Sorbonne in Paris. It was the late 1870s and a stupendous time to be a scientist. The secrets of the atom and the elements were starting to ripen and if you worked hard and were extremely clever, one or two or more might be yours to pluck to yield a Nobel Prize.

Mineralogy professor Charles Friedel, who had, by the way, been a student of Louis Pasteur's, encouraged Jacques and Pierre to investigate the pyroelectricity of crystals. What was going on and was it in any way important? Jacques and Pierre began heating one type of crystal after another to see if all, any, or only tourmaline gave off voltage as their temperature changed. The answer they received was almost as confusing as possible: it wasn't just tourmaline, but it wasn't all crystals either.

How to explain it? Why could you heat some crystals and generate electricity and heat others and generate only a hot crystal? There wasn't an obvious answer. Jacques and Pierre racked their brains. What was the thread that united the few types of crystals that were pyroelectric and excluded the vast majority that weren't?

It took some time, many sleepless nights, oft-stroked beards, and much pacing back and forth across the laboratory, but the breakthrough finally came. Whether or not a crystal was pyroelectric boiled down to that crystal's symmetry. Or, rather, it was related to that crystal's certain lack thereof.

At that time, crystallography was pre-pubescent, although it had been around since the early 1600s. X-rays, which are indispensable for discerning the atomic structure of crystals, would not be discovered until 1895 and the X-ray diffractometer, crystallography's favorite tool, would not be invented until about 10 years after that. Amazingly, when you consider the depth of their insight, Jacques and Pierre could have had but little information about the molecular structure and composition of the crystals at their disposal.

Now we know now that if you categorize crystals by their axes of rotation and their planes of reflection,[4] you wind up with 32 different possible types of crystals. These crystal classes, as they are called, have names like triclinic-pinacoidal, gyroidal, and ditrigonal-dipyramidal and have anything from no symmetry to four threefold axes of rotation. Ten of these crystal classes are pyroelectric and none of those is what is called centrosymmetric. In other words, no crystal whose faces have equivalent opposite faces are pyroelectric.

That centrosymmetric crystals were incapable of pyroelectricity, this much Jacques and Pierre were able to grasp, even if the bigger crystallographic picture was unavailable to them.

Confusingly, there are more than ten noncentrosymmetric crystal classes but only ten crystal classes capable of pyroelectricity, so the case for pyroelectricity

---

[4]They are planes of reflection because we are talking about three-dimensional objects. In two dimensions, these would be axes of reflection.

4.3 Piezoelectricity

turns out to be special. Pyroelectricity requires crystals that are not just noncentrosymmetric but inherently polar (having their positive charges occurring more toward one side and their negative charges occurring more toward the other). The electric effect of this only becomes clear, however, when a disturbance like heating wipes away the surface charges that compensate for the internal polarization of the crystal.

But, again, we're jumping way ahead of Jacques and Pierre. Because they didn't have information about the arrangement of atoms in their different crystals, they didn't know about the polarity of pyroelectric crystals. But they were bright enough to realize two things. The first was that pyroelectricity was little more than a cabinet curiosity; it was time to move on to bigger and better things. And the second was simply sheer brilliance. They made the leap it was to realize that compression or expansion of noncentrosymmetric crystals along a (noncentral) axis of symmetry[5] should produce electricity. They figured that if pyroelectricity was possible, it was possible that pressure-generated electricity was possible too.

Did the brothers Curie sit down, stunned by that moment when it all clicks, the confusion clears, and you know you've got it right? That moment when it all makes such sense you have no idea how no one in the entire history of humankind realized it before? Or did they hop straight up and start testing it? It was a monumental challenge. If they were to put pressure on certain crystals, could they generate a voltage across them? If so, would they be able to measure it? It was time for more brain-wracking. How could they definitively demonstrate that there was such a thing as what they came to call piezoelectricity?[6]

They were the right chaps for the job. Definite MacGyvers. Designing and building previously undreamt of equipment was one of their considerable fortes. To prove piezoelectricity, they cut careful slabs from crystals of table sugar, tourmaline, topaz, quartz, and potassium sodium tartrate (aka Rochelle salts, a laxative as well as a chemical widely useful in chemical reactions and therefore easily available in chemistry labs). They placed the small crystal slabs between two pieces of foil and this they placed in a vice. Jacques and Pierre connected a voltmeter to the two faces of the crystal slab and when they tightened or loosened the vice thereby changing the pressure on their various noncentrosymmetric crystals, volts, hurrah!

It sounds easy, doesn't it? But their apparatus was built exactly as it needed to be and the selection of crystals and the cuts they made of them could not have been better planned in terms of achieving a response (if they'd tested the wrong crystals or even the wrong axis of the right crystals, the result would have been a big zip, zero, zilch). Jacques and Pierre published their first paper on the topic, *Développement par pression de l'électricite polaire dans les cristaux hémièdres à faces inclinées*, in 1880, when Jacques was twenty-five and Pierre was twenty-one. Quartz and the laxative, by the way, displayed the strongest piezoelectric effects.

---

[5] In terms of the crystal faces, not in terms of the arrangement of atoms.
[6] That is, pressure-generated electricity.

The year after the Curie brothers published their discovery of piezoelectricity, a fellow denizen of the Sorbonne, Professor Gabriel Lippmann (who would in 1893 hire a one Marie Skłodowska to work in his laboratory), published calculations indicating the existence of the reverse. Voltage applied to a piezoelectric crystal should cause it to contract or expand depending on the polarity of the voltage.

Theory is one thing, physical measurements another. The brothers Curie leapt into action. This one was really going to be a challenge. The change in the size of the piezoelectric crystal was going to be superlatively slight. Even if they hit that crystal with a whopping 4400 volts of electricity, they could expect an expansion of less than nine nanometers which is, as they cheerfully point out in one of their publications, one fiftieth of the wavelength of visible light (of the color violet, whose wavelength range tops out at 450 nanometers). At best they could expect a change in size there was no way they were ever going to be able to see.

So they said (probably not in these exact words and not just because they were speaking French), what the hell, let's just invent a better way to measure pressure. Because if we pen the crystal in, preventing it from expanding when we pass a voltage through it, it should exert a precisely predictable amount of pressure on its enclosure. We just need to measure the pressure the crystal exerts on its enclosure. And what better pressure detector than a quartz crystal itself, which we shall hook up to an electrometer.

What they did was put their quartz in a vertical vice that they were then able to apply electricity to, causing the quartz to try to expand. The pressure this pushing exerted on the bronze plates of the vice was passed down through the bottom bronze plate to the three quartz crystals beneath it which, in response to this change in pressure, gave off electricity that could be measured by the electrometer. The upper and lower units of this contraption, of course, had to be completely electrically isolated from each other. Other complications had to be dealt with as well (they were trying to measure a tiny change in pressure on a system that was itself exerting a lot of pressure). But it was another splendid brothers Curie apparatus. It worked. When the upper quartz crystal, confined within the vice, was subjected to electricity, the quartz crystals below the vice recorded an increase in pressure in the column of material on top of them. Jacques and Pierre Curie published these results confirming Lippmann's prediction of the reverse piezoelectric effect within a single year of Lippmann's publication of that prediction.

Do you see what the Curie brothers did there? They used piezoelectric quartz crystals to great effect in a piece of electrical equipment. Did they have any idea what a giant leap for mankind this was? Almost 150 years on, we still use piezoelectric quartz crystals at the heart of much of the tools and technology most people on Earth use not merely everyday but almost all the time. Piezoelectricless, modern life would be a bit steampunk. So it baffles the mind that Jacques and Pierre Curie (perhaps alongside instigator Charles Friedel or reverse piezoelectric effect theorist Gabriel Lippmann) did not receive a Nobel Prize for their discovery of piezoelectricity and innovation of the first technology built around it. Perhaps it took too long for the rest of the world to catch on to the possibilities provided by piezoelectric crystals of quartz (never mind sensitive pressure sensors, piezoelectric

4.3 Piezoelectricity

quartz would come to mean sonar, telecommunications, consumer electronics, and wrist watches both affordable and exact). But by then it was too late for a Nobel Prize. Pierre had died in 1906, Charles Friedel in 1899, and Gabriel Lippmann in 1921. Only Jacques, who hung in there until 1941, might have lived long enough for the light to dawn in the Nobel Committee's minds, but if that light had ever dawned, nothing was ever done with it.

4.4 Sonar

It's pleasingly fiendishly clever what people have managed to make of the piezoelectric powers of quartz. Take the production and detection of ultrasound. We probably couldn't do it without a piezoelectric crystal like quartz.

This story begins during the opening years of the First World War, a few decades after the invention of the first two devices employing piezoelectricity, namely the manometer Jacques and Pierre used to prove the reverse piezoelectric effect and a radiation detector that aided Pierre and Marie during their discovery and isolation of several of the first known radioactive elements. The problem this time was the North Atlantic, or to be more specific, icebergs and submarines. Mainly submarines.

Chemistry and physics were not the only disciplines that had been developing at lightning speed during the late 1800s to early 1900s. Warfare was also rapidly modernizing. Submarines had only first been used to torpedo enemy craft in 1864, during the blockade of Charleston, South Carolina, by the Confederacy's hand-propelled H.L. Hunley (which got too close to its victim, the USS Housatonic, and was also lost in the blast). Now in 1915, they were a big thing in the North Atlantic. Germany, having declared the seas around Great Britain and Ireland a war zone, was using its U-boats to sink ships, mainly military and merchant, although it did not consider passenger ships entirely off limits. Hence the RMS Lusitania, on May 7, 1915, 11 miles off the south coast of Ireland and steaming for Liverpool, torpedoed out of the blue (and to tremendous explosive effect). The Lusitania disappeared entirely beneath the waves within 18 minutes and within a span of time not so much greater than that, 1200 of the 1962 people on board died of physical trauma, drowning, or hypothermia.

In addition to inspiring international outrage and inching the United States of America closer to entering the First World War, the sinking of the Lusitania by a German submarine frayed edgy nerves. If you were on the water, be you fisherman, merchant, officer, cadet, seaman, or passenger, terrible death could come at any moment from a predator you couldn't see and thus had no chance to avoid or defend yourself against. The situation wasn't sporting, it wasn't moral, and it was absolutely devastating Britain's navy and trade. Some means of detecting the subaquatic menace had to be found. Because if you knew where a submarine was and in which direction it was heading, you could bomb it, torpedo it, or just plain old head away from it as fast as you possibly could.

Originally, the plan was to listen for the drone of diesel engines of submarines traveling at the surface or the noise of the propellers and random clangs and clongs of submarines cruising underwater. Active echolocation (the production of sound meant to bounce back off an object, relaying information about is existence, location, and trajectory) was a futurist fantasy, if even that (it took some time for the concept of echolocation to creep up at all onto the radar of human consciousness). In the early years of the twentieth century, we hadn't yet coined the word and we were decades away from realizing that the squeaks, shrieks, clicks, and songs of bats, whales, dolphin, and shrews were animals having beaten us to the echolocation punch by millions of years.[7] But the idea that you could passively locate vessels by listening for them (and the slinking of their anchor chains, the clomping of feet on deck, and, who knows, maybe even the shantying of sailors) was old, harking back to at least Leonardo da Vinci, who, having spent his life thinking up everything that didn't yet exist, had made some late fifteenth century sketches of a listening tube that could be stuck into the sea.

And so by the early years of the twentieth century, ships had long been equipped to listen for the noise of dangerous objects. Undersea bells helped, placed at the bases of fixed features it would be bad to bump into in the dark of night, storm, or fog. But listening for chimes was no good as a means to avoid moving things (like other vessels and icebergs) unless you could hang them all with bells like Alpine cows. Such chimes had the additional disadvantage of not relaying precise information about the location (or movement) of the obstacle (or enemy).

The first active echosounder was the Fessenden oscillator, which Canada sent out into the North Atlantic in 1915 on its submarines. The Fessenden oscillator worked like a loudspeaker, emitting a sound at low frequency (about 1 kilohertz, or 1000 cycles per second) and therefore long wavelength (about 1 meter, or 3 feet). Unfortunately, at such a wavelength, the sound spreads outward more or less in all directions, thinning its energy out over a greater and greater area as it travels away from its source. Such a broad beam becomes quickly too weak for its reflections to be detectable and it gives only imprecise information about objects smaller than, say, huge icebergs and the seafloor (that is if it detects them at all). Also a loudspeaker is a great way to announce your existence to the enemy, especially if you enemy is, as they all were at that time, generously kitted out with hydrophones. If this was going to be the reality of echolocation, there weren't going to be many takers.

However, at about this time there was a man who grasped that ultrasonic beams were the way to go. Unlike an undisciplined sonic beam, an ultrasonic beam would spread as a cone of relatively small angle at its apex (perhaps a single degree), making it precise, directable, and nondissipative. Perfect for echolocation. The man with this plan was Constantin Chilowsky, a Russian living in Switzerland who had

---

[7]In 1944, when Dr. Donald Griffin and his student Robert Galambos presented concrete evidence for echolocation in bats (and coined the term), they simply shocked the field of animal physiology, although the idea that bats navigated using sound had been lurking on the lonely, more than half-forgotten fringes of science since the 1700s.

4.4 Sonar

noted that a beam of sound at 50 to 200 kilohertz would travel though seawater at 1500 meters per second, giving it a wavelength somewhere between 0.7 and 3 cm. In addition, he proposed that this beam of ultrasound would reflect off any object it encountered and because you knew the speed of the ultrasound, the time it would take for its round trip back to the detectors could be used to calculate the location of the object causing the reflection.

Constantin sent his proposal for producing ultrasound using a magnetic method to the powers-that-be in France, where he thought it might stand some chance of being developed. The authorities passed it on to Paul Langevin, a professor of physics at the Collège de France whose expertise included magnetism and electrodynamics. By sheer coincidence, Paul had been a doctoral student of Pierre Curie and briefly, several years after Pierre died in the unfortunate coincidence of busy boulevard, rain-slicked cobbles, his head, and the wheels of a horse drawn carriage, the lover of Marie.[8] Thus Paul happened to be one of the few people alive on Earth at that time intimately familiar with piezoelectricity and its reverse and how best to coax the effects out of a crystal of quartz.

Paul and Constatin worked together in the south of France to design a submarine detecting device. Two of them, actually. One was based on Constantin's magnetic means of producing ultrasound, and the other, which was Paul's idea, used an electrostatic transducer to convert electrical energy into the physical energy of ultrasound. These submarine detecting devices, patented in 1916 and 1917, didn't quite have what it took to be useful, especially in a theater of war, but they introduced the world to the basic principles of sonar.

Perhaps unsatisfied, Paul went on to ponder the piezoelectric effect and whether or not you could use quartz crystals to generate and receive ultrasonic sounds. (Over in Britain, on behalf of the Royal Navy's Board of Invention and Research, Lord Rutherford was also giving this a go, but he failed by virtue of sticking to a cut of crystal more suited to sensitive detection than high energy ultrasonic emission. Meanwhile in America, Walter Guyton Cady was using Rochelle salts to produce ultrasound.) It took some time to sort out (although considerably less than the 21 years and 17 days it took the US Patent Office to accept the patent), but on June 21, 1920, Paul Langevin submitted US patent 2,248,870: PIEZOELECTRIC SIGNALING APPARATUS[9] in which he added quartz to the device that he and Constantin had designed. In this patent, he proposed subjecting quartz cut to just the right dimensions to produce an electric oscillation. Through the reverse piezoelectric effect, this would cause the quartz to repeatedly expand and contract, doing so at a frequency both high and controllable. In direct contact with seawater, this high frequency vibration of the quartz would translate into the ultrasonic beam described in Paul's earlier patents with Constantin. Further, the ultrasonic waves

---

[8]Later, Paul would become the grandfather of a boy who would go on to marry one of the granddaughters of Pierre and Marie, as if it they were all part of an epic scientific soap opera.

[9]When the patent was finally granted, in 1941, Paul Langevin was 69 years old and truly fervently resisting fascists in occupied France. He died in 1946 having lived, at least, to see the end of World War II.

returned by reflection off an object like a submarine would, through the piezoelectric effect, induce the quartz crystals to emit electric oscillations that would be picked up by the receiver, allowing for calculation of the travel time of the ultrasonic waves and thus position of the object reflecting the signal.

This really now was what we would call sonar, although it would take more research and development to refine the technique. Sonar, especially with respect to detecting submarines, only became finally truly workable during World War II. Later, sonar leaked into civilian and scientific life, revolutionizing fishing, navigating, finding wrecks, and mapping the seafloor. Over the many decades that followed, people found further uses for ultrasound generated by the expansion and contraction of piezoelectric quartz crystals (like stupendously expensive toothbrushes and the generation of fuzzy images of your child-to-be). Nonetheless, it is easy to think of this year (1920) as marking the time when the use of the piezoelectric effect began to spread far and wide beyond the confines of the Curie laboratory, itself little more than a shed in the midst of the great metropolis of Paris.

4.5 Quartz Oscillators

Of course that doesn't mean that before Paul Langevin's 1920 patent application no one had known about the piezoelectric effect. Since its discovery at the end of the 1870s, the piezoelectric effect had been presented by the various Curies and their associates in patents and other publications and discussed at scientific meetings. While Paul Langevin was developing sonar, other engineers and physicists were beginning to wonder to what neat uses they might be able to put small, clean piezoelectric crystals of quartz. One such person was Walter Guyton Cady, professor of physics of Wesleyan University in Middlefield, Connecticut and researcher working with General Electric and the US Navy. At some point in his playing around with the piezoelectric effect, he hooked a quartz crystal up to a variable frequency oscillator. This led to an unexpected discovery. Quartz would vibrate with high amplitude at a frequency close to that of the natural resonance of the crystal, but basically not at all at other frequencies.

This meant two things, both of which Walter quickly grasped. Such a resonator could amplify a signal and it could serve as a frequency filter or stabilizer. Once you'd determined the resonant frequency of a particular crystal (or the resonant frequencies of a set of crystals of different lengths and widths), you could use the crystal or crystals to calibrate radio or other high frequency circuits to a high degree of precision. Alternatively, you could create a current of precise frequency over a slightly broader range of input frequencies that approached the natural resonant frequency of the crystal, something that you could control by the way you cut the quartz crystal. These features were and still are critical to telecommunications because they are used to stabilize the frequency with which telephone calls (and

4.5 Quartz Oscillators

now also cell phone calls) are transmitted, preventing them from mingling with other calls. Likewise, crystal resonators allow the frequencies that the phone does not wish to receive to be filtered out at the receiving end of the phone call.

Following Walter's discovery, such stabilizing and filtering was also almost immediately put to use by radio stations tired of having their signals drift into those of the stations on neighboring frequency bands.

Having sent that patent in April of 1923, it only took Walter only until October to submit the next blockbuster. A crystal resonator could also be used as a crystal oscillator. The oscillating (that is, alternating) electrical current applied to a quartz crystal would cause it to alternately deform (expand or contract) and then relax back toward normal, as with Paul Langevin's use of the piezoelectric effect to produce ultrasound. Walter took this a step further by noting that the relaxation back toward normal would produce voltage. The time that this took was related to the natural resonance of the crystal, which, again, could be controlled by the way it was cut. Thus, quartz crystals could be used to generate electrical signals of fixed, precise, and precisely selected frequency. The added bonus of quartz is that it wouldn't be strongly affected by environmental conditions like temperature, small shifts in pressure (related to altitude or weather), or humidity. Quartz also does not rust, does not get brittle, is difficult to crack or otherwise damage, and very little energy is needed to keep it resonating. Forget coiled springs and pendulums, then. Quartz oscillators were the way you wanted to keep time.

The way that works, at least for a quartz clock, is the quartz crystal is cut and the oscillating voltage on it is set so that the quartz vibrates at roughly 33 kiloherz (33,000 cycles per second), which puts it a bit above the range of normal human hearing (although the original quartz clock, invented by Warren A. Marrison of Bell Telephone Laboratories in 1927 and accurate to an at that time mind-blowing tenth of a second per day, ran at 100 kHz). In the case of a mechanical timepiece, the resulting pulses of current produced by the quartz crystal are used to drive the motor that turns the mechanism of the clock. This motor is of the synchronous type, meaning that the number of revolutions of its shaft is precisely pegged to the cycling of the current. For a clock or watch with no mechanical moving parts, the cycling current produced by the quartz crystal drives a digital counter.

Following the invention of the transistor and the integrated circuit, quartz clocks could be reduced to wristwatch size, sending the Swiss watch industry into a panic and near death spiral in the 1960s (before they hopped on the crystal bandwagon with their iconic Swatch). These days, timepieces are no longer the things of kings and the tower of the town church. They're tiny, hundreds of times more precise, and no longer cost a fortune. Not only will you not keep yours carefully oiled, level, and wound so it can be handed down through generations of your descendants, you'll probably just toss it when the battery wears out. Unless you're going in for the shiny status of a luxury watch, buying a new watch will be cheaper than having the battery changed.

4.6 But Why Is There a Piezoelectric Effect?

At this point in the story, one question remains unanswered; one, in fact, that remained unanswered for some time after the discovery of the piezoelectric effect and its first implementations into technology. Why is quartz piezoelectric? We've said it is related to a certain type of lack of symmetry of quartz crystals and that it hinges on their having opposite faces which are not equivalent. But that's not an answer, so much as a pattern observed. The actual answer must be in atomic terms, namely that of imbalance in the placement of positively versus negatively charged atoms within the quartz crystal. Why does electricity result when the crystal is compressed (or allowed to expand) along certain axes but not others?

Here is where the conceptual unit cell comes in handy. Although the quartz crystal itself grows through the addition of silica tetrahedra, we can "build" an idealized quartz crystal by bringing three unit cells together to form a hexagonal prism. If we were to look straight down at this idealized quartz crystal, we'd see the hexagon informally sketched in Fig. 4.5. But what Fig. 4.5 also shows is that along some but not all axes of symmetry for the crystal there is asymmetry in terms of atoms and charge.

This is also true of the unit cell. If you were to bisect one unit cell with such as done for one unit cell in the middle panel of Fig. 4.5, the two resulting halves are not mirror images. One half of the cell contains four oxygen atoms while the other contains two, although the two halves have essentially the same number of silicon atoms. Because the silicon atoms are positively charged (containing an excess of protons in the nucleus versus electrons in orbit around the nucleus) and the oxygen atoms are negatively charged, there is a charge imbalance.

Such imbalance is also true of the hexagonal crystal if you approach it the right way. If you were to bisect the hexagon in Fig. 4.5 by running a line from the lower left hand corner to the upper right hand corner, you'd end up with two halves that

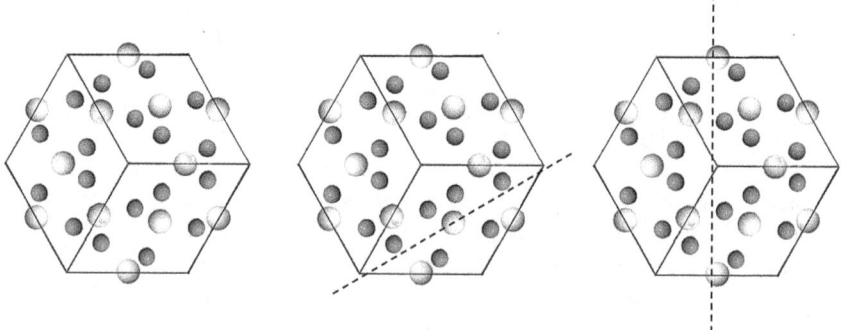

Fig. 4.5 An idealized sketch, based on the unit cell from the *left panel* of Fig. 4.4, looking down on three unit cells of quartz together forming the hexagonal shape of a crystal of quartz. Silicon atoms are shown in *white* and oxygen atoms are shown in *grey*. The *dashed lines* bisecting a unit cell in the *middle panel* and the hexagon in the *third panel* are discussed in Sect. 4.6 of this chapter

are perfect mirror images. The same is true for the hexagon of Fig. 4.5 bisected along its horizontal "equator." But if, as in the third panel of Fig. 4.5, you were to bisect the hexagon by running a line from the middle of the top horizontal border through the middle of the bottom horizontal border, you would not end up with two equivalent halves. The left half contains fewer oxygen atoms than the right half, but the number of silicon atoms is the same. If you were to push those horizontal faces closer together, you would reduce the spatial separation of charges inequivalently in the left and right halves of the crystal. Thus you would generate voltage. However, if you were to press it together along an axis of symmetry in terms of charge, the balanced charges would remain balanced and no electricity would result.

Lastly, one of the reasons that the piezoelectric effect is as great in α-quartz as it is, is that roomy crystal structure leaves some room for the silica tetrahedra to rotate, allowing for even greater change in the dipole moment during compression or expansion along the piezoelectric axis than would occur with a more rigid framework.

So New Agers don't have it all wrong. Crystals of silica (and a lot of other materials, including that laxative) do indeed have energetic powers. Interestingly, though, the rise in New Age beliefs, especially the ones involving mystical crystals, *postdate* the discovery of the piezoelectric effect.

It is perhaps all about what you do with observations of natural phenomena. Study them systematically and then figure out what cool things you can build around them (like manometers, sonar, telecommunications, radio, and pretty unprecedentedly precise timekeepers) or put them in a velvet pouch and worship them. One's good for Nobel prizes, fortune-building patents, and the satisfaction of changing the way that people live and the things that they can experience and do. The other sparkles some and perhaps brings inner peace. Both approaches are very human and how you judge one versus the other depends on how you feel about technological progress and our ever increasing dominion of the universe around us.

Whatever the case, we invite you to momentarily unite the worlds of science and technology and of alchemy and New Age spirituality. Grab a crystal of quartz and give it a squeeze. You won't feel it, but you'll know it's there, that fabulous, world-changing power of piezoelectricity.

Further Reading

Duck F (2009) 'The electrical expansion of quartz' by Jacques and Pierre Curie. Ultrasound 17:197–203

Langevin P (1941) Piezoelectric signaling apparatus. US Patent 2,248,870. Submitted 21 June 192, issued 8 July 8 1941

Manbachi A, Cobbold RSC (2011) Development and application of piezoelectric materials for ultrasound generation and detection. Ultrasound 19:187–196

Marrison W A (1948) The evolution of the quartz crystal clock. Bell Syst Tech J 27:510–588

McWhan D (2012) Sand and Silicon. Oxford University Press, Oxford

Chapter 5
Glass Houses and Nanotechnology

Wouldn't it be cool to live in a world that was full of microscopic houses made of glass? Not just plain glass houses either, but ones with nanoscale details, the minuscule evocations of windows and doors, flagpoles, antennas, and weathervanes. What about living in a world that also boasted bazillions of organisms roaming around with glass skeletons or glass plates of armor and contained land plants woven through with glass shards. But you already know the punchline. There are no *what ifs* about this. Our world *is* full of microscopic glass structures wrought in purposeful and epic detail. Thank silica biomineralization. Creatures lurking on almost every major branch of the eukaryotic family tree (sketched in Fig. 5.1) have been producing biominerals of that amorphous, hydrated silica known as opal for as long as there have been such things on Earth as animals, if not for a few hundred million years longer. Today, to name but a bunch, diatoms, sponges, radiolarians, choanoflagellates, chrysophytes, euglyphids, ebridians, heliozoans, thaumatomonads, horsetails, grasses (including rice, wheat, and bamboo), reeds, rushes, sedges, palm trees, forget-me-nots, maize, squashes, bananas, sausage trees, arrowroot, and orchids biomineralize silica and the oceans, rivers, lakes, and their sediments; soils; dust; and many of the fruits, vegetables, and grains we eat abound with this biomineralized opal.

"What about us?" you might be wanting to shout. "We also make glass. Doesn't that count?"

Yes, we do make glass, but not in a way that can be classified as biomineralization. For mineral production to qualify as biomineralization, it has to take place inside of a biological cell and the resulting product must become integrated into the body of an organism. Also, compared to nature's natural glassmakers, our efforts at glassmaking clock in somewhere between crude and embarrassing.

To give us credit, though, we haven't had 750 million years to tinker with the techniques and, now that we're on to them, we won't need 750 million years to catch up. This isn't because we are so supremely clever. If we manage to master

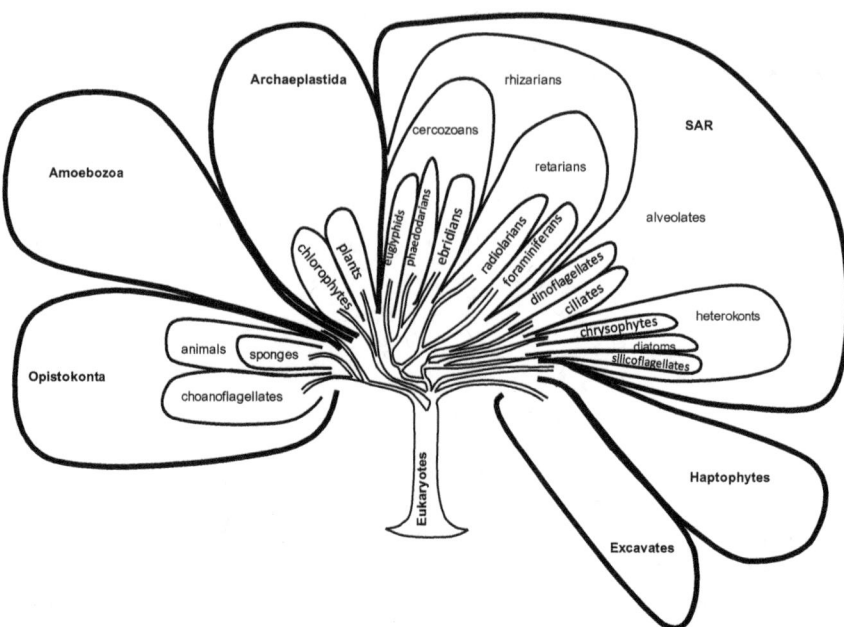

Fig. 5.1 A rough sketch (not to scale in terms of evolutionary distances) showing the locations of various silica biomineralizing groups on the eukaryotic family tree. The largest lobes represent what are known as the different "super groups" of eukaryotes. While the ability to biomineralize silica occurs within all super groups, not all the groups within the super groups are capable of it; hence there are many phyla (etc.) of eukaryotes missing from this tree. Also the branches and lobes of this tree as it is pictured, in addition to not being to scale in terms of evolutionary distance between organisms, represent a work in progress rather than scientific consensus set in stone

silica biomineralization techniques (at least in laboratory or industrial settings, not as a supremely magnificent biohack)[1] it will be more because we're not averse to murderous industrial espionage. Some of our best materials scientists and engineers, molecular biologists and geneticists, and nanotechnologists have already set themselves on the job of popping open silica biomineralizing diatoms, land plants, and sponges and probing them for their molecular and genetic glassmaking secrets. So someday soon we too will produce glass from water at room temperature with speed, efficiency and sub-microscale magnificence. In fact, we've already figured out the basic basics of this. But we have a long way to go before our nanotechnological glassmaking efforts amount to anything half as impressive as what gets made naturally.

Because we're being inspired by biomineralization, this story of our current to future nanotechnological glassmaking skills begins billions of years ago, back at the dawn of biomineralization (and of silica biomineralization in particular).

---

[1] Although we would be up for that.

5.1 Silica-Centric Musings on the Origin of Biomineralization

Once upon a time, and for a very long time after that, all life was soft. Not only was everyone spineless, there were no bones at all. The killing of one organism by another to steal its CHNOPS[2] had to be carried out without teeth, claws, horns, tusks, hooves, or antlers. But, hey, no problem, there were also no scales, no shells, or no other hard parts that required sustained effort or special tools to breach.

Though idyllic you could not say it was, it was at least unescalated. What food web there was had few trophic levels and churned slowly. Many cells dined exclusively on the post-primordial soup of dissolved organic matter exuded by other living cells or produced by dying cells, for example, when viruses swept through, leaving droves of microorganisms bursting in their wake. Those cells that required solid food relied upon phagocytosis, that slow process whereby a single cell, such as an amoeba, engulfs another, such as a bacterium, and digests it whole.

In time, soft-bodied, single-celled predators upped the ante by developing pseudopodia (which are like tentacles) and membrane-ringed oral grooves (which are like mouths that become stomachs). This gave them considerably greater power to capture (yay, pseudopodia!) and to enclose prey (yay, oral groove!) in a segment of membrane that was pinched inwards into the cell and, thus closed, used as a digestive vesicle. But as effective as these innovations were (and still are, for they are still in widespread use), they still merely consisted of soft parts.

Things carried on thusly toothless and unarmored for billions of years and might have remained forever so had the ocean not been saturated with several of its salts. That's what happens when solutes are always flowing in following their weathering out of rocks but there's no independent mechanism for their removal. Eventually things get so thick, certain salts precipitate on their own.

Because the bulk of the Earth's crust is silicate rock, as it weathers, a lot of dissolved silica gets added to the ocean. So, before biomineralization, there was plenty of silica there in its dissolved form, silicic acid, atomically symbolically speaking, $Si(OH)_4$ and also commonly referred to as dissolved silica.

These single $Si(OH)_4$ units, or monomers, of silicic acid are simply individual, independent silica tetrahedra whose four negatively charged corner oxygen atoms each have a positively charged hydrogen ion bound to them, as introduced in Chap. 1 and shown in Fig. 1.3. But when concentrations of monomeric silicic acid get too high, meaning things get crowded enough for the monomers to frequently bump into each other, the monomers start to combine to form dimers, trimers, and, eventually, polymers of silicic acid.

---

[2]Carbon, hydrogen, nitrogen, oxygen, phosphorus, and sulfur, the major constituents of the organic matter which makes up living things.

This was most certainly true of the ocean that prevailed during the first *three or so billion years* of Earth history, before silica biomineralization existed to provide a prodigious output of solid silica from the ocean. In the oceans of the Precambrian,[3] concentrations of dissolved silica built up to such excess that monomers of silicic acid were driven to bind together to form dimers of silicic acid, which are two silica tetrahedra joined together, one losing a hydrogen (H) and the other losing a hydroxyl (OH) to come to share one of the corner oxygen atoms: $(HO)_3Si-O-Si(OH)_3$. The formation of dimers was followed by the formation of trimers and then an increasingly crazy jumble of polymers of silicic acid that, through this increasing formation of silicon–oxygen–silicon bridges between increasingly networked tetrahedra, grew into granular silica gels. The granules of silica continued to grow, now through a process, known as Ostwald ripening,[4] that favors the growth of large granules at the expense of smaller ones, to mature into solids.

Concentrations of silicic acid in the Precambrian ocean were even high enough for silicic acid to diffuse through seafloor sediments to become incorporated into solid silica cementing itself into massive bedded cherts forming in deeper strata. Likewise, concentrations of dissolved silica were also high enough for silica to precipitate on the surfaces of iron oxides that themselves had been precipitated as a side effect of the activities of photosynthetic bacteria. This iron-precipitated silica accumulated on the seafloor and contributed to the formation of some of the spectacular banded iron formations (BIFs) of the Precambrian (they would have been more accurately named banded iron-silica formations, especially as they contain much more silica than iron).

All of these processes entailed silica precipitation out of the silica supersaturated Precambrian ocean, no intentional biological intervention required. And what an amazing thing it must have been—an ocean so full of dissolved silica, it all but rained glass onto the ocean floor.

So would conditions have remained until today but for the fact that, as any old school crystallographer with a big, bushy beard could tell you, precipitation from a supersaturated solution goes faster if you provide surfaces suitable for nucleation. Such surfaces invite the adsorption of specific solutes onto them or, just more generally, lower the energetic barrier to precipitation that exists even in supersaturated situations. In the ancient ocean, there were a number of abundant materials

---

[3]The geologic eon known as the Precambrian began with the final formation of the Earth 4.6 billion years ago and ended 541 million years ago in an explosive diversification of animal life and, incidentally, of mineralized shells and skeletons.

[4]During Ostwald ripening, smaller colloids of silica would tend to dissolve faster than they would have had additional silicic acid units aggregating on to them. But larger colloids (with their more favorable surface-to-volume ratio) would have had silicic acid units added faster than they would lose them. Thus over time smaller colloids would become more and more rare as their silica tetrahedra increasingly accumulated onto the larger colloids, eventually growing the larger colloids up into the size range we term particulate.

5.1 Silica-Centric Musings on the Origin of Biomineralization

(in addition to those accidental particles of iron oxide) that encouraged the precipitation of silica. There were proteins and other organic compounds that had been biologically synthesized for various cellular tasks but also happened to bind and precipitate silica. The existence of such compounds inside of single-celled organisms soaking in a supersaturated ocean may have made intracellular silica biomineralization an all but inevitable development.

In fact, avoiding unwanted intracellular silica precipitation in this conjunction of circumstances could have been difficult.[5] One biochemical, osmotic, or metabolic misstep and blam! Mineralization. Poor cell. Did one big clod of a mineral making sudden violent space for itself in the cytoplasm or was it lots and lots of teeny little ones solidifying themselves into place around the cell membrane? Either way... What such sudden unwanted mineral blob or blobs didn't prove fatal would have at least been awkward, adding the equivalent of cement shoes to cells that needed to stay up in sunlit surface waters or gumming up the physical operation of the cell.

Over time, failures would have died out, leaving only cells that had either gotten better at preventing intracellular mineral precipitation or had learned how to manage it so it didn't kill them. This learning how to manage it meant confining precipitation to specific localities within the cell, transporting the resulting mineral towards the outer edge of the cell and then extruding the mineral out into the environment. You can think of that as bagging up and taking out the trash or as the basics of biomineralization. When cells crossed that fine line, what began as waste management became actual, intended intracellular biomineralization that produced usually crystalline solids[6] of specific size, shape, and thickness that became integrated into the cell itself and fundamental to its identity and way of life.

So when was that? When did true biomineralization (siliceous and otherwise) begin to occur?

In the grand scheme of things, relatively recently. While life may have been well established and diverse enough to include photosynthesizers by 3.8 billion years ago, unambiguous evidence for any type of biomineralization is currently no older than about 800 million years. In other words, true biomineralization popped up sometime around 80% of the way through life's current total history. And while we'd love to tell you that the first biominerals were made of silica, the oldest biominerals that we can say for sure we know the composition of were phosphatic. But other fossils of similar age may have contained biomineralized silica, although the evidence for this remains inferential.

---

[5] The same was also true for calcium carbonate minerals like calcite (found today in clams and foraminiferans, for example) and apatite (found today in corals and pteropods, among other things).

[6] The exception to this is of course biogenic silica, which by virtue of the way it has condensed out of solution (i.e., fairly chaotically) is way too disordered to be considered crystalline.

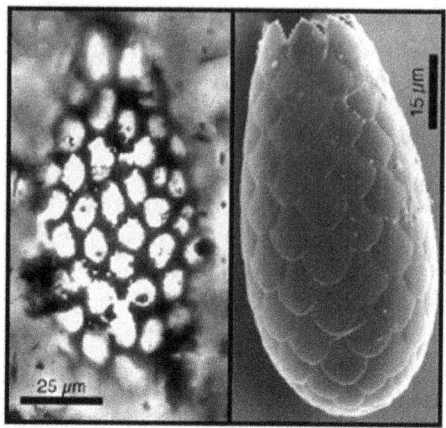

Fig. 5.2 The *vase-shaped* microfossil *Melicerion poikilon* is shown on the *left*. As with the modern testate amoeba *Euglypha tuberculata*, shown on the *right*, its multiplicity of empty spaces may have originally hosted silica scales. These images are reprinted from Porter (2006). The Proterozoic fossil record of heterotrophic fossil eukaryotes. In: Neoproterozoic Geobiology and Paleobiology (Eds. S Xiao and AJ Kaufmann), Springer, pp 1–21, with permission

5.2 The Early Fossil Record of Silica Biomineralization

The current candidates for the first true biominerals of silica are scales that were present on unicellular plankton that lived approximately 770–740 million years ago. The remains of these ancient plankton are known as vase-shaped microfossils. One of them is pictured in the left panel of Fig. 5.2.

Before you get too excited, we have to confess that these remnants that are now called vase-shaped microfossils were not originally made of silica; they consisted of rigid, biomineralized structures made out of organic matter.[7] But, still, these vase-shaped microfossils are incredibly informative. For one, they collectively indicate that not only was biomineralization being carried out by several different species of eukaryote at this point,[8] it was being carried out by species in two very different major clades (that is, phyla) of eukaryotes and that is a pretty big deal.

Careful digestion of the metamorphosed sedimentary rocks in which the vase-shaped fossils are embedded yields three-dimensional casts of the robust outer structure of the cells. Based on the details of their structure, some of the vase-shaped microfossils clearly fall into the phylum Amoebozoa (which has come to include lobose amoebae and two of the three main types of slime mold), but

---

[7]If you're having trouble imagining a biomineral made of organic matter, think of things like tortoise shells, horses' hooves, or your own fingernails.

[8]Eukaryotes are organisms whose cells contain membrane-bound organelles like a nucleus, mitochondria, and chloroplasts. This encompasses everything that isn't an archaea or a bacteria.

5.2 The Early Fossil Record of Silica Biomineralization

others clearly belong to the phylum Cercozoa (which now counts filose amoebae among its ranks).

Lobose amoebae versus filose amoebae. They don't sound so different. But they are. To put it in perspective, animals are more closely related to amoebozoans than amoebozoans are to cercozoans. That's how far back in time the line that led to amoebozoans (and choanoflagellates, animals, fungi, and slime molds) and the line that led to cercozoans (and a lot of other things, including plants and algae) split apart.

Another thing we know from the vase-shaped microfossils is that these organisms may have been able to create more than one type of biomineral. The fossils are all casts of the rigid organic polymer casing that these extraordinarily distantly related organisms had covered themselves with. That organic casing alone qualifies as biomineralization, but some of the holey polymer casings (such as shown in Fig. 5.2) look like they are molded so as to hold other things in place. For instance, they may have once hosted scales that, not being made of the same material as the organic framework that was preserved, dissolved away long ago. We think that those now-missing scales may have in some cases been made of silica.

The miniature proteinaceous cisterns of some of the cercozoan vase-shaped microfossils are notable this way because these fossils closely resemble modern day euglyphids, a type of cercozoan testate amoebae *that is covered in silica scales*. As evidence goes, this is not a slam dunk, but it opens up the possibility that some of those ancient cercozoans may have also made themselves scales of silica. If this was the case, silica biomineralization was occurring in the oceans as early as 770 million years ago.

By 550 million years ago we have silica biomineralization going on for sure. We know this from the multitude of sponge-made silica that starts to show up in the fossil record. Sponges themselves were pretty new at this point and, as most of them still do today, they contained what can loosely be called a silica skeleton— a somewhat chaotic mesh of different types of spicules (some of which are pictured in Fig. 5.3) that look very similar today to the ones from the fossil record of 550 million years ago.

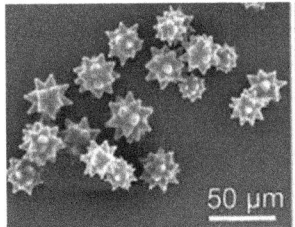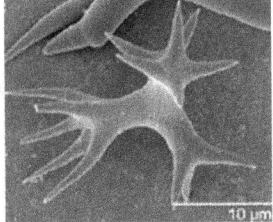

Fig. 5.3 Individual sponge spicules come in various shapes and sizes. These images are reprinted from Cell and Tissue Research, Vol. 328, Maldonado M, Riesgo A, Intra-epithelial spicules in a homosclerophorid sponge, 639–650 (2007) with permission from Springer and from Advances in Marine Biology, Vol. 62, Maldonado M, Ribes M, van Duyl FC, Nutrient fluxes through sponges: Biology, budgets, and ecological implications, 113–182 (2012), with permission from Elsevier

By about 540 million years ago, during what is called the Cambrian explosion, the number of types of different animals (what can be called the diversity of animals) increased, well, explosively. It was also during this time that most of the animal phyla we know today became established. Sponges, which are animals, who had already been around, took this opportunity to really diversify. As both the overall total number of individual sponges and the overall total number of sponge species increased, so did their way with silica biomineralization. They left the traces of this behind in the fossil record as an increasing abundance and variety of the silica spicules that they produced.

At roughly the same time, radiolarians, a group of amoeba-like protists that produce skeletons of silica, also made their first appearance on this Earth and, like sponges and, possibly, some of those cercozoans, began extracting dissolved silica from seawater and turning it into biominerals made of amorphous silica.

Altogether this blossoming of silica biomineralization meant that the cycling of silica within the seas was now and for the next several hundred million years controlled not by inorganic chemical processes related to supersaturation but by silica biomineralization carried out (mainly) by sponges and radiolarians.

5.3 Not All Biomineralization Is Silica Biomineralization

We should flag at this point that our enthusiasm for silica may be leading you astray. Not all biomineralization nor even most biomineralization is silica biomineralization. While this silica biomineralization we've been describing was developing over evolutionary time in fits and starts, it was not the only biomienralization game in town. The biomineralization of calcite, of aragonite, of apatite, and of many other minerals was also spinning up during this long span of time. In fact, it is the blooming of biomineralization in general to some degree enabled the impressive expansion in animal diversity that occurred during the Cambrian explosion which marks the beginning of our current geologic eon (the Phanerozoic), the one which is characterized in no small part by the existence of animals.

By the time the Earth had gotten through the Cambrian explosion, animals had really picked up the biomineralization ball and started to run. Brachiopods and their precursors had popped up making little shells out of calcium carbonate and calcium phosphate. Trilobites and arthropods had arrived on the scene sporting exoskeletons of chitin, calcium carbonate, and calcium phosphate. The world's first molluscs and echinoderms showed up making shells and tests of calcium carbonate. In addition to this, biomineralization within that unfairly grouped group of unicellular critters that is protists, which had been dabbling with biomineralization since the beginning of biomineralization, also continued developing.

By the time this heady biomineralization revolution had calmed down, animals and protists had in total learned how to make, incorporate, and use more than 60 different types of biominerals. And it was not the case that one organism figured it out and then biomineralization spread out down the line of descendants. While that of course occurred, it was more the case that it occurred *also*. Biomineralization, even of the same mineral, was invented independently many dozens of different times by different animals, protists, land plants, and even bacteria.

5.4 The World's First Arms Race

The intense grow in of biomineralization from all but nothing in the late Precambrian to increasingly widespread in the early Phanerozoic can be cast, not too fancifully, as the world's first arms race. Perhaps in response to the deployment of pseudopodia and oral grooves and certainly because there was a certain futility to being eaten before passing on genes, some single-celled organisms, such as those vase-shaped cercozoans and amoebozoans, learned to armor themselves with the biominerals they had accidentally turned out to be able to make. This worked perhaps brilliantly at first, cutting down predation rates by forcing predators to work harder for their supper. But as these things inevitably go, eventually the innovative shielding resulted in innovations from the other team: biomineralized teeth, claws, and other tools useful for breaking through mineralized defenses. And what response must have come of that? More robustly biomineralized defenses, also spikier and more all encasing. In turn, thicker bones, sharper teeth, bigger claws, and so on would have turned up amongst the predators.

There would have been all sorts of side effects to this arms race. One was that hard parts and the diverse array of body parts that organisms started to make with them made organisms capable of inhabiting ecological niches and habitats that hadn't even existed before. Animals with skeletons could also get bigger and start doing things, like walking, scuttling, and swimming, that nothing on Earth had ever done before. Motility meant that they were more energetic and thus needed more food more often. The food web began spinning faster as there came to be a greater number of motile predators armed with biominerals and every innovation in biomineralization that broke the stalemate between predator and prey contributed to making the food web more violently voracious. It would take hundreds of millions of years, but eventually innovations in biomineralization gave way to the conscious and intentional production of tools, weapons, and technology for capturing (or raising) and then devouring prey, in part starting but most certainly not ending with those stone tools made of silica presented in Chap. 3.

As with all arms races, civil uses trickled out of the militant developments. Mineralized skeletons (internal or external) turned out to be great as leverage for muscles and structural support, allowing for complex, three-dimensional body plans that withstood gravity in or out of the water. (Land ho!) To put it mildly, being able to make hard parts opened up new directions in which animals (and plants) could evolve and new environments they could move into.

So while silica biomineralization cannot beat its chest as it points out how lousy the world would be without giraffes, it can look proudly upon plains of tall, swishing grass and amber waves of grain, every blade of which is thick and tough with bits of biogenic silica, and think, I helped with that.

5.5 How to Make a Glass House: Man Versus Nature

5.5.1 Man

Just because vertebrates don't biomineralize silica, it doesn't mean that they don't all not make glass. As you may have pointed out earlier, we people have been up to it for a while. Several thousand years, in fact, in what might possibly be the only biological glassmaking (possibly) initially inspired by the torturous tubes of fused glass produced when lightning strikes sand (fulgurites, they're called, in case you are interested).

What are the techniques that we use to make glass?

Like lightning, when we humans make glass, we start with solid silica, specifically something like quartz sand. We must melt this quartz into a viscous lava before we can work with it and this requires heating it to just about 2000 °C (3600 °F). Such hot heating by human hands takes not just a furnace or a kiln, but a furnace or a kiln built to withstand such extreme temperatures (most aren't).

These temperatures are not just difficult to safely contain, they are expensive in terms of fuel. Luckily, we have found there is more than one way to melt quartz. Adding additives to the pile of powdered quartz can lower the melting point down into the neighborhood of 1500 °C (2700 °F). Sodium carbonate (which your grand- or great-grandparents may have called washing soda or soda ash) works well for this, although the sodium silicate produced in the ensuing fusion reaction will dissolve in water faster than the Wicked Witch of the West sublimated away.[9] Happily, this can be countered with calcium oxide (aka quicklime). Adding it to the melange to be melted prevents sodium silicate from forming. Sometimes we throw in other additives too, like aluminum oxide, iron oxide, copper oxide, or lead oxide, to add color or sparkle to the end-product glass or to make it easier to cut.

Once whatever siliceous mix we've made is molten, we can shape it. One technique involves pressing a gob[10] of molten glass at the end of a blow tube into a mold made of wood, metal, or clay. This imprints a pattern on the outer surface of the glass. The imprinted gob is removed from the mold and inflated at the same time its tip is pulled. Pulling straight merely elongates the gob to the intended size of the finished vessel. Simultaneously twisting spirals the pattern imprinted on the gob. This

---

[9]Although she did scream, "I'm melting!", which is not the same thing as sublimating (going directly from solid to vapor), which is what it looked like she did. (Obviously, she was not a chemist.)

[10]Indeed, that is the technical term.

kind of glasswork has been going on since at least the ancient Egyptians, which is somehow no surprise given that they lived at the edge of that veritable ocean of quartz sand that is the Saharan Desert (and which gets hit by lighting from time to time).

A gob of molten glass can also be mold-blown. This means it is slightly inflated, put into a mold, and then further inflated to fill in the nooks and crannies of the design. The glass is not removed until it has cooled enough to solidify. For thousands of years we have made all sorts of stuff this way, too. Ordinary urns, jars, drinking glasses, and other such vessels, but also vessels in spectacular shapes such as human heads so detailed as to have curls of hair, curves of nostril, and eyelids. The molds can even be reused for repeated production of the same object.

These days, where low-cost, high-throughput trumps unique and artisan, bottles and jars are usually the products of two molds whose products are sealed together to form the hollow container, leaving visible seams.[11] Meanwhile, large sheets of flat glass, such as for windows and display cases, are made by floating the molten glass out over a molten metal like tin.

But the process is not yet over. Once our glass creations have cooled, unless we are cheapskates, they need to be annealed. Annealing entails heating the solidified glass back up to 550 °C (1020 °F) (for soda-lime glass) or 1200 °C (2200 °F) (for quartz glass) for a few hours and then letting it cool back slowly down to room temperature. This gentle heating followed by slow cooling allows stress points produced by the stretching, twisting, thinning, and bending involved in the shaping of the viscous glass to relax. At annealing temperatures, the glass becomes just soft enough for the silica tetrahedra to align themselves in a better, stronger way, slightly more orderly way, thereby eliminating local weak points in the glass.

Despite the difficulty and energetic expense, humanity has been making glass by working molten silica for about 5500 years. The finest things we have to show for this work? Any list is arguable, but here goes ours: the 13th century stained glass interior of Sainte Chapelle in Paris; the glass flowers, diseased fruits, and marine invertebrates made by the Blaschkas, many of which may be seen in the Harvard Museum of Natural History; chandeliers; millefiori and Murano glass objects; thousands of microscopic diatoms arranged into microscopic mandalas on microscope slides by artists like Johann Diedrich Möller; the Crystal Palace of London created by Joseph Paxton; all the soaring glass-and-iron or glass-and-steel train sheds of rail stations around the world; glass curtain-walled skyscrapers; and disco balls. There is no denying that all of these things are magnificent. But their magnificence generally either comes from their massive size or resplendent colors. Very little of it impresses with the virtuosity of its shaping or the magnificence of its detail. And none of it impresses on the nanoscale.

---

[11]Not to go in for the popular sport of hipster-bashing, but Mason jars, the favorite beverage receptacle of otherwise artisan-loving hipsters, are mass produced using this technique.

5.5.2 Nature

Silica biomineralizers use entirely different techniques to make glass.

To begin with, silica biminералizers don't start with solid quartz, they start with dissolved silica. Thus they don't have to gather a fairly pure quartz sand and they don't have to melt quartz. Instead their first task is to take monomers of silicic acid up out of water and transport them across a cell membrane, into the interior of their cell (or cells).

Extracting dissolved silica from water is not a terribly difficult task. It just takes the right chemical. Silica biomineralizers take in silica tetrahedra using proteins known as transporters. These transporters specifically handle silica, which they often co-transport in along with an ion like sodium, and are embedded through the outer cell membrane such that they serve as a tunnel through the membrane (Fig. 5.4). To put it roughly: the end of the transporter that grabs silica faces the external environment and the end that lets go of it faces the inside of the cell.

Although the job of every silica transporter is the same—to bring silica into the cell—there are quite a number of different silica transporters. In part this is because some silica transporters evolved entirely independently of all the others, and in part this is because transporters that got passed on from generation to generation changed as genetic mutations piled up over time. There is thus not just a variety of transporters among different types of silica biomineralizing organisms (which can be as different as sponges and redwood trees), a single silica biomineralizing cell has a small arsenal of different transporters at its disposal—one for almost every occasion.

One of the ways that silica transporters are characterized by chemists is by their ability to bind and transport dissolved silica (a measurement that, while requiring some expertise to make, is millions of times easier than discerning the structure or composition of a transport protein). Some of the transporters have what we call a

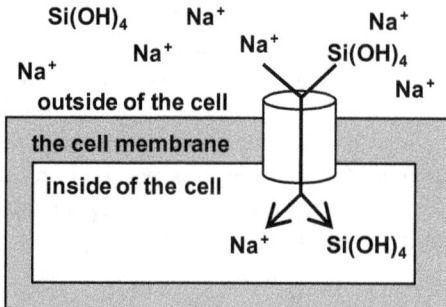

Fig. 5.4 Conceptualized sketch of a membrane-spanning silica transporter in diatoms. Transport of silicic acid through the transporter from outside of the cell to inside of the cell is coupled to the passage, also through the transporter, of sodium ions that is driven by the gradient in sodium concentrations (higher outside than cell than in)

high affinity for silicic acid. This means that this transporter is extremely good at binding to dissolved silica and can operate at its maximal rate at very low concentrations of dissolved silica. The advantages of this for organisms whose growth depends on dissolved silica should be obvious.

But silica biomineralizers have low affinity transporters as well. These require high concentrations of dissolved silica in order to operate, because they are not as good at binding dissolved silica. This sounds stupid until you realize that high-affinity transporters have a serious drawback. If they're running at top speed at low concentrations, they're not running any faster at high concentrations. And top speed is nothing like the speed a low affinity transporter can transport silica into a cell at when dissolved silica concentrations are high. Where and when dissolved silica concentrations are abundant, easily maxed out high-affinity transporters are doing the equivalent (if sort of the opposite) of bailing out a boat with a teaspoon.

In other words, it makes sense to have a suite of options.

Transporting silica into a cell almost always involves moving it against (rather than with) a concentration gradient (concentrations of silica are higher inside the cell than outside the cell). Thus it is work to pump silica into a cell and work requires energy. But cells hate to expend energy they don't absolutely have to. That energy could go towards growth and cell division, or as we like to think of it, dividing and conquering. Instead of expending their own energy, silica biomineralizing cells commonly couple silica uptake to the uptake of ions spontaneously running "downhill" towards the area of lower concentration inside the cell.

In diatoms, the world's greatest phytoplankton (IOHO), silica transporters take advantage of the imbalance between sodium ion concentrations outside the cell (much higher) and concentrations inside the cell (much lower). As the lack of equilibrium drives sodium ions through the silica transporter, one to two silica tetrahedra are pulled through per sodium ion, depending on the transporter.

The place where the case may be different is with vascular plants.[12] Their silica transporters only appear to operate when fed cellular energy.

The next thing all true silica biomineralizers must do during the process of silica bimineralization is move silica through the cell cytoplasm so that it ends up either at the site of biomineral deposition or inside membrane-bound vesicles where silica is stored until it is time for opal precipitation to begin. In practice, a little bit of both goes on. While silica structures won't start forming until the cell has accumulated some minimal amount of silica necessary for their construction, once biomineralization is progressing, concurrently uptaken silica directly supplements the construction.

One intriguing thing about the buildup of silica inside a cell before biomineralization is that so much can accumulate within that tiny volume of the cell (or even tinier total volume of the storage vesicle), silica saturation is exceeded. Silica

---

[12]Vascular plants are those land plants with conduits made of xylem, a type of tissue reinforced with lignin, for conducting water from root to everywhere else.

colloids, gels, or even solids should automatically condense. But they don't, at least not until the cell decides that it is the proper time to begin biomineralization.

At this point, silica biomineralizers guide the condensation of silica into a solid of precisely predetermined shape. This shape is determined by the protein- and carbohydrate-rich template that the biomineralizer has built to induce precipitation in certain places and prevent it in others. The biomineralizer can also use different proteins and carbohydrates in different orders and ratios to control the size of the spheres of silica that precipitate in different locations and set the stage for further growth of the biomineral. This guiding organic template winds up woven through the final silica mineral and in this way has additional influence on the strength, solubility, and texture of the end-product opal.

Using this set of techniques, silica biomineralizing organisms create a diverse array of structures and fabrics. They can make silica that is solid or silica that is riddled with pores set out in swirling grids of geometric precision. They can create flat sheets of silica or silica that curves. They can produce walls of silica that are, at least by microscopic reckoning, thick or thin. When the mood strikes, they can add protuberances that look like spikes, merlons, or the teeth of a zipper. And they can produce silica that easily dissolves and is therefore disposable or they can produce silica so recalcitrant it sinks to the sediments after the death of the organism and survives on for hundreds of millions of years.

5.6 Some Silica Biomineralizing Organisms that We Are Learning From

Precipitating silica out of solution at room temperature sure beats having to melt quartz. How are we learning to do this? By studying silica biomineralizing organisms. What follows are some of the things we've learned about what some of these organisms can do and therefore what we someday might be able to accomplish.

5.6.1 *Choanoflagellates*

One group of organisms being tapped for its silica biomineralizing secrets are the choanoflagellates. They are ancient, unicellular, and cute, sort of like microscopic tadpoles (although the choanoflagellate doesn't have a tail, it has a flagella). Truly cool choanoflagellates form a fine basket of silica, called a lorica, around themselves. One such silica lorica is pictured in Fig. 5.5.

To give you a bit of backstory on the choanoflagellates, they have been around metaphorically forever. Genetic analyses say that they and the creature that would give rise to all animals split off from a common ancestor. That makes

5.6 Some Silica Biomineralizing Organisms that We Are Learning From

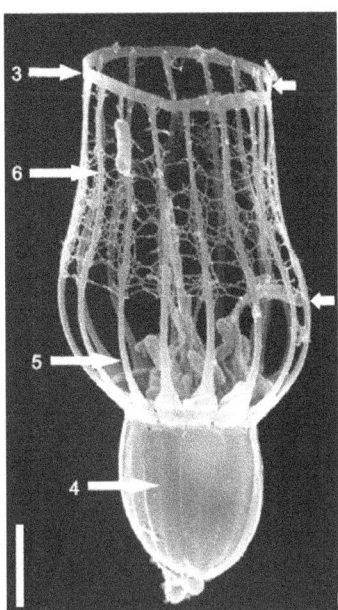

Fig. 5.5 Scanning electron microscope (SEM) image of the silica lorica of the choanoflagellate *Volkanus costatus*. *Arrow* 3 points toward one of the transverse costae. *Arrows* 4 through 6 point out some of the vertical costae. The two *unnumbered arrows* demark the upper and lower extent of this lorica's web-like veil. The *bar* included to give you a sense for the truly tiny size of it all is 2 μm long. This image has been reprinted from Protist 161, Leadbeater BSC, Choanoflagellate lorica construction and assembly: The tectiform condition. *Volkanus costatus* (=*Diplotheca costata*), 160–176, (2010), with permission from Elsevier

choanoflagellates the one and only sister group to animals. If you can imagine the eukaryotic family tree (or better yet, have another gander at Fig. 5.1), choanoflagellates and animals are two, paired terminal[13] twigs on the lobe of the tree that is opisthokonts (animals, choanoflagellates, fungi, and a bunch of different single-celled critters, many of which are parasites). We're also pretty sure that the immediate ancestor to choanoflagellates and all animals looked an awful lot like a choanoflagellate. (Which makes it unsurprising that sponges, which were one of the world's first animals, contain some cells (called choanocytes) that are the spitting image of choanoflagellates.)

Not all choanoflagellates have a silica lorica, but that of those that do is made of strips of silica, known as costae, that are up to 3 millionths of a meter long (also known as 3 microns or μm) and 100 billionths of a meter thick (also known as 100 nanometers or nm). These lorica can be so elaborate as to contain 300 costae arranged in two layers. The outer layer of costae runs longitudinally down the

---

[13]Those that do not lead to others.

egg-shaped organism, while the inner layer of costae, which holds the outer layer in place, is more helical or latitudinal in nature.

Even a fairly simple silica lorica, such as the one pictured in Fig. 5.5 is exquisite enough to make you wonder how a microscopic, unicellular organism manages, without a brain or hands or anything like that, to assemble such a small, complex, and exact glass structure.

The assembly of the lorica begins within a parent cell that is about to divide itself into two daughters, each of which will need a lorica. In one of the two main types of choanoflagellate species, the daughter cells emerge naked (without a lorica), find a safe place to settle, and then go about building their loricas themselves. In the other type, the parent cell constructs a lorica around each daughter while it still holds them within itself.

How does such a tectiform choanoflagellate parent cell, as they are called, accomplish this? It builds within itself protein-rich templates upon which silica condenses, resulting in silica segments of precise length and width. The cell uses these segments to construct costae that it then accumulates them into bundles. Again, this is all occurring within the microscopic confines of the parent cell.

Some of the bundles of costae are laid out longitudinally and others latitudinally and this is no accident. Longitudinally oriented bundles will become the longitudinal costae of the outer layer of the lorica of each daughter cell. Horizontally oriented bundles are destined for the construction of the helical or latitudinal costae of the inner layer of each daughter cell.

When there are sufficient costae for two lorica, cell division commences and this is the end of the parent cell. The parent choanoflagellate divides the material within itself into two even portions (it's only fair). But before the parent cell releases these two daughters, it flips and shoves them within itself so that they bump into the bundles of costae that have accumulated. The little tentacles of the choanoflagellate cell help this process along. Thus the two daughter cells are able to acquire the microscopic bundles of costae that are needed for their loricas and it is only once they have them all that they are they kicked out of the meager, tattered remains of the parent cell to fend for themselves and to finish constructing their loricas.

Around each of these newly independent daughters, the bundled costae begin to arrange themselves into the two interlocking layers of the lorica. This requires quite a symphony of movement, one which is in part conducted by tiny tentacles, lasts from five to fifteen minutes, and, incredibly, generally goes off without a hitch no matter how complex the design of the lorica. No crunches, crashes, tangles, or ill-placements occur, even when 300 individual costae are involved.

A simple lorica, just a gridded basket, is the easiest sort to arrange. But there are choanoflagellates that have costae that wind helically against their vertical costae. These choanoflagellatees must rotate these long costae around themselves while sequentially attaching them to the other costae in the framework. It is a bit like spinning a microscopic glass cocoon where each fiber must find its exact preplanned place down to the nearest nanometer.

There is no *thinking* going on when this happens. This is a unicellular organism! Somehow this complex construction is conducted without consciousness. If we could figure out how to create a set of instructions that could carry themselves out like that, we too could create self-building glass nanostructures (or perhaps even self-building macroscale glass buildings).

Unfortunately, while it's easy to understand how genes can code for certain chemical sequences, such as those which make up the templates for silicification, how genes carry the information for the physical assembly of the solid pieces of silica remains for the moment beyond our understanding. But by combining genetic work with computer modeling; close examination of choanoflagellate silica, its structure, and its properties; and the processes by which lorica are constructed, we're working on sorting it out.

5.6.2 Siliceous Sponges

For an animal that during its adult life sits anchored to the seafloor and filters food out of the water day after day after day after day, sponges command fiercely loyal attention and enthusiastically heartfelt admiration from a small army of ecologists, paleontologists, biochemists, taxonomists, materials engineers, and biogeochemists, among others, including those who'd like to build better silica.

Perhaps surprisingly, this simple sponge lifestyle is a highly successful one. Sponges have been around, abundant, and living that sponge lifestyle for more than half a billion years and, unlike a lot of other organisms out there in this ecologically cataclysmal Anthropocene epoch, they don't appear to be heading towards extinction anytime soon. Although obviously, we wouldn't be around to collect our winnings, we'd happily bet that sponge will still be filtering along long after *Homo sapiens* has disappeared from this Earth (one way or another). In the meantime, like the choanoflagellates, sponges are teaching us a thing or two about making glass.

This is because sponges are not just sessile survivors, they are avid biomineralizers. Somewhere between many and most sponges (in other words, all of the hexactinellids and many of the demosponges) produce silica and not at all half-heartedly. These siliceous sponges fill themselves up with biomineralized silica to the point where their silica can be called a skeleton and that seems like a gross understatement because while its individual elements may be spindly, there are so many of them packed in there, silica within a siliceous sponge is far from sparse.

As a result of this avid silicification, altogether each year marine sponges produce approximately 20 million metric tons of unfastened spicules, fused skeletons, and stalks of silica. In so doing, they cement considerable new mass into ocean reefs made from opaline glass. And this does not take into account any of the silica made by freshwater sponges.

Sponges are multicellular animals and this more or less means that they are made up of genetically identical but differentiated cells that must work together individually or bundled into tissues. The successful sponge lifestyle depends in part upon the work of choanocytes, which are cells that look like choanoflagellates and like the last common ancestor between choanoflagellates and all animals. Like choanoflagellates, each choanocyte has a flagella which it can whip. Together, the flagella-whipping of all the choanocytes in a sponge creates a current of water that flows into the sponge through its pores, travels through the body of the sponge (which is full of internal cavities and canals like a... sponge), and exits from an opening at the top of the sponge. The water drawn in brings with it dissolved silica and dissolved organic matter that may be taken up by sponge cells. The current also carries in particles of food (living and detrital) that may be filtered out and phagocytized, especially by the choanocytes.

The outflow of water expels all the garbage.

The spongy material that makes up the inner portions of the sponge is known as mesohyl. It is made of fibrous proteins, like collagen and spongin, and it is what you have in your hand when you buy a natural bath sponge that is not a loofah (which, by the by, is an edible cucumber).[14] In siliceous sponges, the mesohyl contains shards, slivers, and other skeletal elements of glass that contribute significantly to the sponge's robustness and its unpalatability. As an additional line of defense, a carpet of all these tiny, sharp slivers shed on the seafloor around the settled sponge discourages the approach of benthic organisms, like sea urchins, that find sponges tasty (or at least edible).

The silica pieces that sponges make, some of which are shown in Fig. 5.3, are fantastic: stars, hooks, jacks, needles, and yawning medieval medical/torture device-ish things, all made of shining, transparent glass. Some of these spicules are microscopic. Most you can just see with the naked eye (at least until you are on the latter side of middle-age). Others are long and skinny and may be longer than you are tall despite having been extruded by one microscopic cell.

How does a siliceous sponge manage to make all these spicules of different shapes and size (and, again, without hands or a brain to guide the process along)?

The production of the more modestly sized silica spicules begins in sclerocytes, which are amoeba-like cells that wander freely within the sponge. When a sclerocyte sets itself to producing a spicule, first it constructs a filament, mainly out of the protein silicatein. It does this within a subcellular vesicle called the silicalemma. Because silicatein acts as a silica-precipitating enzyme, silica pumped into the silicalemma polymerizes around the silicatein filament and grows from there. Once the spicule has completely silicified, the sclerocyte extrudes it into place in the mesohyl and, until it is time to make the next spicule, its job is done.

Parts too large to be constructed inside of a single cell require a different approach. In this case, the filament around which silica precipitates is extruded from the sclerocyte to the mesohyl before any silica has condensed on it. Release of dissolved silica and additional silicatein into the vicinity of the extruded filament then drives the deposition of silica on the filament in this extracellular area inside the body of the sponge.

Using these techniques, the most siliceous siliceous sponges fill their mesohyl with a finely filigreed, three-dimensional mesh of glass fibers. But that's not all.

---

[14]No joke.

5.6 Some Silica Biomineralizing Organisms that We Are Learning From

Some of the glass sponges that live in the deep sea set their highly siliceous bodies on top of a silica stalk that makes it possible for the sponge to be both anchored to the seafloor and raised up into the slightly faster flow of water that exists above the seabed. It isn't unusual for these stalks to consist of a bundle of biomineralized glass fibers a meter or more long.

The super longest silica stalks (that we know of) are produced by the glass sponge *Monorhaphis chuni*. These sponges live at the bottom of the deep sea. The first *M. chuni* to see the light of day was dredged up during that mid to late nineteenth century rush to lay deep sea cables and explore the deep, dark innards of the ocean, in this case during the German Valdivia expedition undertaken in the Atlantic and Indian Oceans in 1898–1899. Jaws must have dropped at the sight of it. Today we would say the stalk looks like a bundle of fiber optic cables. Each individual spicule in the bundle is 8 millimeters (5/16ths of an inch) in diameter and up to 3 meters (10 feet) tall.

This remains an unrivaled feat of silica bioineralization.

The people most currently obsessed with the monumental basal spicules of deep sea sponges are materials scientists. How does a sponge make silica that is so strong and yet so flexible and if we could figure it out, could we do it too?

Both the strength and the flexibility of the long stalk spicules comes from their basic design, which is multilayered and organically interbedded. These long spicules consist of concentric layers of silica up to 10 μm thick. Between each two layers of silica is one layer of organic matter, namely the organic matter that was used as the template to catalyze silica precipitation. Without the resulting flexibility plus mechanical strength, the stalks of deep sea glass sponges would not withstand the currents and occasional disturbances of life at the bottom of the deep ocean.

And withstand they can. One spicule in a stalk collected from a *live* sponge on the 1100 meter (3600 foot) deep floor of the East China Sea has been estimated to be 11,000 years old. Bear in mind that there is no way to patch or repair a spicule once it has been produced.

In other words, that is glass that lasts.

And we've taken note. The alternating application of solutions of silica-precipitating proteins and solutions of silicic acid could be used to artificially produce similarly multilayered and organically interbedded silica structures of both great strength and flexibility. All sorts of structures could be possible; spheres or fibers would merely be the easiest to accomplish. Not to rampantly speculate, but glass bricks, glass sculptures, glass panes, and glass jewelry produced to specifications and without the need for a furnace are four possibilities that quickly spring to mind.

5.6.3 Diatoms

Do you know diatoms? They are phytoplanktons that live in the ocean, rivers, ponds, lakes, and estuaries. As a group, they currently carry out 20% of the Earth's

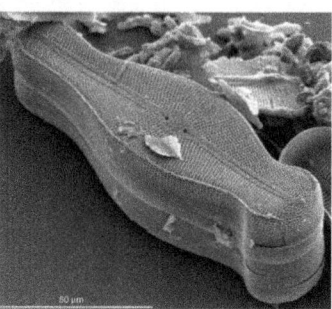

Fig. 5.6 Scanning electron microscope (SEM) image of the frustule of the freshwater pennate diatom *Didymosphenia geminata*. Both the pores on the valve faces and the girdle bands holding the two valves together are clearly visible. This image has been reprinted from Trends in Biotechnology 27, Gordon R, Losic D, Tiffany MA, Nagy SS, Sterrenburg FAS, The Glass Menagerie: diatoms for novel applications in nanotechnology, 116–127, (2009), with permission from Elsevier

photosynthesis (despite being only about 2% of the Earth's photosynthetic biomass). If you've gone swimming in wild waters, you've certainly swallowed thousands of diatoms or at least gotten them up your nose or in your ears. In a sense, you've also almost certainly eaten diatoms; they serve as the base of the short food chains that lead to big fish. On top of all this, diatoms are unbeatable biomineralizers of silica. As a group, they produce most of the biogenic silica that forms in the world today in aquatic systems and more biogenic silica than all land plants put together manage to do. Individually, they produce a silicified cell wall[15] of unrivaled detail. It is for this last reason that diatoms are the silica biomineralizer that we have investigated most intensively.

The cell wall (officially called their frustule) that diatoms construct out of silica is often described as a pill box or a petri dish and if you have a glance at Fig. 5.6 you can start to see why. The diatom frustule consists of a top half and a bottom half. Each of these two valves, as they are called, consists of a flat face and a vertical rim running around its perimeter. When the bottom valve and top valve are placed together, their side walls overlap slightly, in the manner of a petri dish and its lid.

Diatoms secure their two valves together by encircling the overlapping area with a girdle of lightly silicified bands. Some of these are visible in Fig. 5.6. Aside from hold top and bottom valves together, these girdle bands provide the diatom with versatility. Because glass is rigid and, once constructed, the two silica valves cannot grow; adding girdle bands is one way the cell can add size to its outer covering if it needs to (although generally diatoms stick to living within the space allotted them by the generous size of the frustule). The lightly silicified girdles are also easily

---

[15]The cell wall is a protective layer that sits to the exterior of the cell membrane in the cells of diatoms, plants, and a number of other organisms (but not animals).

5.6 Some Silica Biomineralizing Organisms that We Are Learning From

disposed of when the cell needs to separate its two valves and divide in half, giving one valve to each of its daughters who then go on to biomineralize a new second valve and the girdle bands to hold them in place.

However, calling the diatom frustule a petri dish sells it seriously short. Petri dishes are boring while frustules can be embellished with all manner of intricate glass structures, like multitudes of microscopic pores, vertiginously reaching spikes, and funny little processes that serve purposes we can only at this point ponder.

One silica structure we can at least see one or two points of are pores. Diatom valve faces are riddled with hundreds to thousands of them regularly sized, shaped, spaced, clustered, and aligned. It is, in a word, astounding, but at the same time not surprising. A diatom cell must remain in contact with its liquid medium. That's where it gets the nutrients and vitamins it needs to survive and the carbon dioxide it needs to photosynthesize. These materials need to arrive at the cell membrane in order to be taken up into the cell. Likewise, the cell must expel oxygen and other waste products. But uptake and excretion would be impossible if the cell membrane was contained within a box of solid glass. Thus every diatom frustule is porous.[16]

The poster child for diatom pores is *Coscinodiscus*. Nobody does it better. The members of this genus of giant centric diatoms don't just have pores that run straight through their silica frustule. The valve faces of their frustules consist of multiple layers of silica that have become fused together. And each layer has its own specific set of pores set into it.

Generally, the outer surface of a *Coscinodiscus* valve face hosts hundreds of clusters of pores. Each one of the pores is fantastically tiny—40 nanometers in diameter—and placed so precisely, each cluster of pores looks like a tiny paw print (if paws had up to seven pads). Underneath (and, again, firmly fused to) this surface layer of silica is a second layer of silica. It has pores that, at up to 200 nanometers in diameter, are a bit bigger. The third and innermost layer of silica in the frustule holds the biggest pores. In some species, they may be as much as 1200 nanometers in diameter.

Some sense for all this can be picked up from Fig. 5.7, which shows both the entire inner surface of the valve of one species of *Coscinodiscus* and a close up that includes the sight of the pores in the interior layer of silica visible through the surficial pores.

It is a brilliant design. The necessary gases and dissolved nutrients are able to diffuse from the water the diatom is living in, through the smallest pores, then through the medium-sized pores, and finally through the largest pores to reach the cell membrane. The cell can do what it needs to do in order to survive and grow. Yet the diatom remains physically armored against hungry zooplankton and even against viruses (who can't get their injector down deep enough through the pores to reach the cell membrane).

---

[16] Or mesoporous, if you prefer the technical term.

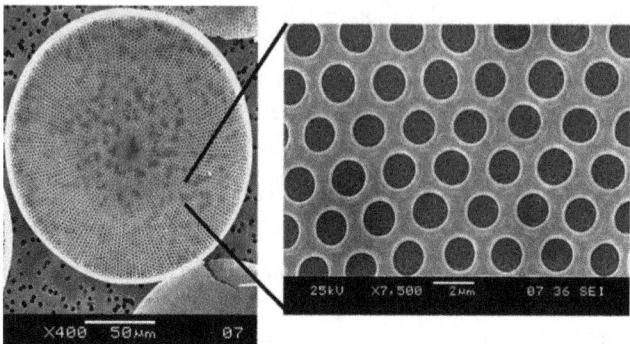

Fig. 5.7 SEM image of the inner surface of a valve of the species *Coscinodiscus wailesii*. The inner surface of the silica contains pores of 500 nanometers in diameter. Visible through them are notably smaller pores, 30 to 40 nanometers in diameter, arranged in clusters in the inner layer of silica. Not shown are the pores on the outer surface of the valve. This figure has been reprinted from Acta Biomateriala 4, De Stefano L, Lamberti A, Rotiroti L, De Stefano M, Interfacing the nanostructured biosilica microshells of the marine diatom *Coscinodiscus wailesii* with biological matter, 126–130, (2008), with permission from Elsevier

But that only settles the more obvious *why* of the porously multilayered structure of this diatom frustule. It doesn't settle the *how*. And it's the *how* that we are interested in as students of biomimetic glassmaking.

As it turns out, the *how* is a bit like the *how* of sponge silica production, but only in the broad sense that dissolved silica is taken up across a cell membrane, delivered to a specific site of silica polymerization, and then precipitated as a solid under the direction of specific organic compounds. For instance, the proteins and polyamines that diatoms use to guide the precipitation of their silica are not the silicateins of sponges (nor are they the as yet not fully identified silica-precipitating polypeptides of choanoflagellates). Instead the diatoms use compounds (silaffins, silacidins, and silica-precipitating polyamines) that are (as far as we currently know) specific to diatoms.

In addition to zooming in on the genetics of silica production by diatoms, we've spent a lot of time working on these silica-precipitating compounds. Because if you manage to extract, isolate, and purify these silica-precipitating compounds, you can do all sorts of fun things. For example, if you sprinkle silaffins into a beaker containing a solution of dissolved silica, nanospheres of silica immediately precipitate. You can control the size of the spheres by using different silaffins or mixtures of silaffins.

Roughly speaking, that means you can turn water into glass whose size you can control at the nanoscale. Not only is this a mighty first step in learning to construct larger but still nanoscale detailed glass structures, such spheres are useful. Chemical functionality can be added to these spheres by adding ions, metals, or larger organic molecules to them as they form, allowing the spheres to act as catalysts for all sorts of reactions (for example, to scrub harmful pollutants from industrial waste waters).

The spheres could also be formed around medicines and given surface properties that would fly under the radar of a person's immune system, thus enabling the spheres to infiltrate into specific tissues of organs before they release, for example, a dose of cancer-killing chemicals. In the future, we'll figure out how to grow slightly larger spheres embedded with pores and that will only enhance diatom-inspired biosilica's drug-delivering utility.

5.7 Siliceous Nanotechnology

Yes, such smooth, cool, easy glassmaking using techniques picked up from choanoflagellates, sponges, diatoms, and other silica biomineralizers is what nanotechnologists call *bio-inspired*. As you may have gathered from our pithy handful of examples, nanotechnologists are totally into it, too, obsessively tinkering with techniques based on biomineralization to make novel, nanopatterned materials in more energy efficient and environmentally friendly ways. If all goes according to their idealistic (and lucrative) plans, these new silica nanomaterials will lead not just to better drug-delivery and more effective catalysts but also to better batteries, more efficient solar cells, improved fuel-, gas-, and contaminant-sorting filters, and various other futuristic micro- and nanoscale devices. If all goes according to their truly most idealistic plan, this siliceous nanotechnology will spare the world some of the current side effects of nearly 10 billion humans living, eating, driving, flying, and being consumer capitalists.

Bio-inspired silica would mean that instead of mining quartz sand and burning lots of fuel to melt it, we could just take water from lakes, rivers, the ocean, or a faucet and add to it the right proteins or polyamines set hierarchically into a template and then... done. The result? A tiny piece of glass patterned precisely to our specifications, embedded with organic matter, and, if we wanted, implanted with electronically or otherwise useful trace elements like titanium and germanium.

We have a ways to go to master it at that level, but we're already pretty good at making pores. By throwing together the right set of proteins and polyamines, we can create self-assembling columns of organic matter that precipitate silica around them such that you end up with a networked honeycomb of silica. Burning the silica honeycomb in a hot oven vaporizes away the organic matter to result in tiny, porous wafers (or spheres) of silica hosting what are officially designated by the International Union of Pure and Applied Chemistry to be micropores (smaller than 2 nanometers), mesopores (2 to 50 nanometers), and macropores (larger than 50 nanometers).

Today what we often use instead of such synthetic mesoporous silica are sedimentary rocks made up of porous diatom frustules, a material mined and then sold as diatomite or diatomaceous earth. The chemical behavior of silica surfaces and all the surface area (and empty space) provided by the porosity of diatomaceous Earth makes it indispensable industrially and even around home and garden. One of the highly porous diatomite's most notable uses is as a stabilizer for nitroglycerin, allowing it to be stored and transported unexplosively.[17] Diatomite is also a useful insecticide; deployed as a powder, it finds its way through the joints of the carapaces of insects where, because its porousness gives it the power to absorb water and lipids, it sucks the bodies of the pests dry. Porous diatom silica is also fabulous as a carrier of detergent in washing powder, as an abrasive in toothpaste and other polishes, as an absorbent for liquids, and as filters for pools, beer, and drinking water.

But there is only so much diatomaceous earth on Earth and it takes work and leaves scars to dig it up out of the ground. In addition, diatomite is not pure. It contains organic matter and clays and other contaminants minerals. Its silica is also somewhat of a grab bag. If you look at it under a microscope you'll see diatoms for sure, but also tons of siliceous sponge spicules and silicoflagellate skeletons. This means that instead of being exact, diatomite's porosity spans a broad range even within a single, small sample. And different diatomites will be different, having been produced at different times and places, under different environmental conditions, and by different consortia of diatoms and other silica biomineralizers.

It would be better to make mesoporous silica to order.

In addition to being able to control the size and frequency of pores, we would, as we've said earlier, be able to add specific functionality to the silica. We can do some of this already. For instance, we already know how to make silica with organic functional groups bound to the surfaces of its tubular pores. These bio-inspired silicas are used as catalysts. We also already know how to embed the organics into the silica itself, making it less likely that the organics will block up the pores, rendering the catalyst clogged and useless.

It's still clearly fairly early days for this field of bio-inspired silica nanotechnology; not only is their great progress still to be easily made, there's undoubtedly lots of money that will come out of many resulting patents. If we two silica scientists had to do it all over again, we might be lured away from biogeochemistry to work on it. Okay, the money, yes, but mostly: who doesn't want to engineer things that might help save the world? But, to let you in on a secret, for us the lure wouldn't be just one of knowledge but also one of artistic and architectural possibilities. Because it would be frankly very cool to build microscopic glass houses (mansions, even!) just like diatoms do. We'd give both Johann Diedrich Möller[18] and Joseph Paxton[19] some serious competition.

---

[17] We call the ensemble dynamite (get it?).

[18] JD Möller was a nineteenth century master of the art of arranging thousands of diatom frustules on microscope slides as fantastic, geometric, microscopic mandalas.

[19] Joseph Paxton spent his life building incredible buildings out of glass and steel. See, for example, the Crystal Palace and the Great Conservatory at Chatsworth, both sadly long destroyed.

Further Reading

Curnow P, Senior L, Knight MJ, Thamatrakoln K, Hildebrand M, Booth PJ (2012) Expression, purification, and reconstitution of a diatom silicon transporter. Biochem 51:3776–3785

Knoll AH (2003) Biomineralization and evolutionary history. Rev Mineral Geochem 54:329–356

Leadbeater BSC, Yu Q-B, Kent J, Stekel DJ (2009) Three-dimensional images of choanoflagellate loricae. Proc R Soc B 276:3–11

Losic D, Rosengarten G, Mitchell JG, Voelcker NH (2006) Pore architecture of diatom frustules: potential nanostructured membranes for molecular and particle separations. J Nanosci Nanotech 6:1–8

Porter SM (2006) The Proterozoic fossil record of heterotrophic eukaryotes. In: Xiao S, Kaufman AJ (eds) Neoproterozoic Geobiology and Paleobiology. Topics in Geobiology, vol 27, Springer, Netherlands, pp 1–21

Wu S-H, Mou C-Y, Lin H-P (2013) Synthesis of mesoporous silica particles. Chem Soc Rev 42:3862–3875

Chapter 6
Chicks Need Silica, Too

Wouldn't you like to have beautiful nails and hair and strong bones? Walk down the supplements aisle of your local drugstore and you might get the idea this takes silica. *Buy these capsules, please!* Colloidal silica gel, horsetail silica, food grade diatomaceous earth, choline-stabilized orthosilicic acid, monomethyl trisilanol. The variety of supplement silica covers it all from dissolved to colloidal to particulate silica and from inorganic to organic forms of silica. Even if you are a skeptic of supplements, the exuberance of offerings is enough to make you wonder whether indeed our bodies need silica. It's a strange idea. We're not like diatoms, glass sponges, choanoflagellates, and many land plants. We don't make glass intracellularly. And silica, unlike iron and other micronutrients, is not incorporated into any of the major enzymes or reactive proteins whose functioning keeps us alive. So why would we have a nutritional need for silica? And why have the silica supplement hawkers dialed down on a specific target: hair, nails, skin, and bones? What do the supplement makers know that most of us don't? Or are they just having us on?

6.1 It's All About the Chicks

For silica supplements on store shelves you can sort of blame 1972. In that year, two scientific papers shocked the world of nutrition by showing that animals (namely chickens and rats) that do not biomineralize silica nonetheless need silica, especially when young and growing.

The chick paper in particular struck a chord and has been remembered down through the years far outside the field of nutrition. Maybe it was the catchy title: *Silicon: as essential element for the chick*. More likely it was that the researcher behind it, Edith Carlisle of the School of Public Health at UCLA, had made

Fig. 6.1 Two one-month-old chicks, one fed properly, the other deprived of silica. This image has been reprinted from Science of the Total Environment 73, EM Carlise, Silicon as a trace nutrient, 95–106, (1988), with permission from Elsevier

investigating silica as a necessary nutrient her life's work. Over the course of her career, she contributed more to our understanding of the role of silica in the body than any other single human being before or since.[1]

In her famous chick experiment, Edith Carlisle took day old hatchlings, used a scalpel to separate them from what remained of their yolk, and then raised them for a month. Half of these chicks were fed a diet devoid of silica. The other half were given 250 milligrams of dissolved silica per kilogram of feed, a diet that thus contained 100 parts per million of the element silicon, to put it in more standard scientific parlance.

The results? Look at the two chicks in the photo that is Fig. 6.1.

One chick looks as a one-month-old chick should. It is, you could say, bright eyed and busy tailed, with well-developed feathers, thick, strong legs, and a comb both plump and perky. The other chick is small, scrawny, and miserable. It's even hanging its head, as if in shame. (In reality, the chick is too weak and deformed for it to be possible to hold its head up.) We'll just let you guess which chick was fed all the silica it needed and which one was denied.

This old photo, which is still famous today (at least amongst silica scientists), launched a thousand scientific investigations. It isn't a stretch to say that at some level, every scientific investigation into the role of silica in the health of skin, hair, and nails, in the formation of bone and connective tissue, in wound healing, and in the reaction of the body to bioceramic implants meant to take the place of bone destroyed by violence or disease can be traced to this photo and the research behind it. It was so shocking. How could something so fundamental to solid rock that further had no obvious biochemical utility to animals (aside from siliceous sponges) be critical for the growth and development of young birds and mammals?

---

[1]Sadly though, while German speakers have seen fit to give her a one paragraph Wikipedia entry, the only other people who seem to remember her today are trying to sell their supplements by linking her fawningly to their products.

6.1 It's All About the Chicks

The first time we two authors saw this photo, we, aside from thinking *Silica rocks!*, were floored by the feat it would have been to keep a chick or a rat or anything else on a silica-free diet. As this book aims to convince you, there is no escape; silica is everywhere. There's silica in the food we eat, in the water, and many other beverages we drink, and it can leach out of glass and pottery bottles, jars, glasses, cups, mugs, and cookware. We also inhale silica, in the form of very tiny rocks or minuscule shards of glass present as the dust in the air, especially if we live a windy desert or work in a glass recycling facility.

Keeping silica-free experimental subjects truly free from silica is a monumental task. You have to house the need-to-be-silica-free subjects in plastic enclosures, keep them in severely filtered air, and give them stringently purified water (bearing in mind that, just to add to the challenge, water distillation units are generally made from glass). You have to feed the creatures a synthetic diet of reagent grade compounds ordered from chemical companies, making sure you've managed to include all the necessary amino acids, trace elements like iron, and vitamins while achieving sufficient calories and a reasonable balance between lipids, carbohydrates, and proteins (or amino acids). But even such a diet would be more *low-silica* than truly *silica-free*. All of the bottled chemicals you need to use contain traces of silica. Because, did we mention, silica is everywhere.

No matter how careful you are, you'll still have to hope that the traces you couldn't eliminate from your silica-free subjects' lives won't be enough to be enough, dietetically speaking.

The second thing the chick photo made us wonder is what made her do it. Why had Edith Carlisle ever thought to wonder what would happen if she deprived a growing animal of silica? It isn't the sort of question that would have come flying to us straight out of the blue and we even have silica on the brain. What offhand remark, idle conversation, or previous set of observations triggered the 1972 investigations?

6.2 Silicosis

Although Edith Carlisle was at the forefront of wondering if silica was an essential nutrient for humans and other animals, she was far from the first to wonder about silica and human health.

Back in mid 1800s to early 1900s, the sense was most squarely that silica was bad for you. Many men worked as miners and miners inhale a lot of rock dust. Those who manage to avoid dying young in cave-ins often instead die prematurely of respiratory disease. One of these diseases is silicosis.

Silicosis is caused by the inhalation of particles of what is loosely referred to as crystalline silica, usually meaning quartz but including non-crystalline forms like diatomaceous earth, industrially fused silica, volcanic ash, and smashed up glass.

This is usually the point in the conversation where the person we're talking to gets wild eyed and worried, thinking about all the time they've spent at the beach, immediately followed by the point where we gently grasp their shoulders and calm them down. Unlike much tinier, much lighter particles of silica dust, grains of sand are too big and heavy for suspension in air except during severe wind. As you are thus not likely to breathe beach sand in, beach sand is no more likely to give you silicosis than is the big crystal of quartz you bought at the New Age shop to keep in a velvet pouch under your pillow to keep yourself electrically healthy. (Confused? See Chap. 4.)

So what happens when silicosis happens?

Again, it all starts when you inhale silica dust. The particles that are not stopped by nose hairs and mucus membranes reach the lungs and some of them lodge in alveoli, those little sacks that occur at the ends of bronchioles. Alveoli, of which you have several hundred million, are critical for respiration. Each alveolus is covered with capillaries and serves as a site for gas exchange: into the blood with the oxygen, out from the blood with the carbon dioxide. When a dust-sized grain of silica wedges itself into an alveolus, the lungs find it irritating. This is an entirely reasonable response, but the cascade that follows is not.

Alerted to the irritation, the body dispatches macrophages, a type of white blood cell, to gobble up the foreign particle. Macrophages may triumph against invasive bacteria and other pathogens, but in silica they more than meet their match. Phagocytized and thus now inside the macrophage, the silica grain does not have the decency to disintegrate. It is after all a small rock. It fails to notice the enzymes, designed to degenerate living matter, that the macrophage launches at it.

Ultimately, instead of the macrophage dissolving the silica particle, the silica particle deeply damages the macrophage. In response, the macrophage begins a frenzied release of protein-degrading enzymes. But instead of destroying the silica, these enzymes digest critical elements of the macrophage. Eventually it dies. Multiply this process by millions and at some point the immune system gets the point. Macrophages aren't working. Time for bigger guns. Cytokines, in other words; cell signaling compounds that play a key role in the body's response to trauma, inflammation, and infection.

As with so many things about the immune response, something which in most circumstances is good and necessary, in other circumstances sets off an escalating spiral of disaster. The release of cytokines against the silica particles triggers the production of collagen, a fibrous protein triple-stranded like 3-ply yarn, in the alveoli. The cytokines simultaneously trigger the release of antibodies against collagen. This amplification leads to further amplification: the development of fibroblasts, the cells whose job it is to produce the collagen that is needed to heal a wound. This is the thing that would be good except for the circumstance.

The circumstance is that this is happening in the lungs. Lungs are a terrible place to weave a web of collagen. As the mats of collagen grow, the lungs become scarred and stiff. Eventually things get so bad, on an X-ray, the lungs look as if infested with tent caterpillar tents and, through a stethoscope, they pop and crackle like a brittle bag.

Someone whose inhalation of silica dust has brought on silicosis that is on its way to being (or already is) the progressive massive fibrosis described above

6.2 Silicosis

wheezes and coughs and struggles to breathe. They feel sharp pain in their chest and have unoxygenatedly bluish lips. The volume to which the lungs can inflate is severely decreased. The lungs also have increased susceptibility to tuberculosis and cancer, among other things.

If the exposure to silica dust has been severe enough in extent and duration, in time the scarred, stiff lungs of this poor soul become fatally ineffective.

It can't be a nice way to go.

Back in those days when a lot of men were miners and dust masks were no more than handkerchiefs, silicosis was not an uncommon way to die (whereas these days, in the US, for example, less than one hundred people die of it each year). This would have been incentive to understand the disease in more detail than cause (inhaled silica dust) and effect (progressively debilitating lung problems ending in death).

One part of better understanding silicosis included figuring out the fate of the silica particles that were inhaled. Did all of them remain in the lungs forever or did some dissolve? Could some of the silica particles be dislodged and, if they made it down into the stomach via the esophagus, would they dissolve in the stomach? What about in the intestines? Finally, if silica did dissolve in the lungs, the stomach, or the intestines, would it make its way into the bloodstream, and thus be circulated throughout the body and perhaps find its way into living tissues other than the lungs?

Not only were these questions pressing in light of silicosis, contemporaneous research by other scientists was revealing that many of the plants we eat are full of silica (stay tuned for the upcoming Chap. 7), thus silica was getting into our gastrointestinal tracts all the time. Wholesale efforts were launched to measure the silica content of different foods as well as of human and animal lungs, livers, kidneys, blood vessels, muscles, blood, urine, feces, and brains.

It was an interesting if messy time to be a silica scientist. The human race was finally discovering that silica was indeed truly everywhere.

6.3 The Dog Days of Silica Medical Research

In the 1930s a one Earl Judson King, medical researcher first at the University of Toronto and later the British Postgraduate Medical School, wanted to know if animals (including humans) could absorb silica through their gut on the one hand and whether or not they could efficiently clear it from their bloodstream (and thus their body) on the other. This leads to some experiments that are, by today's standards, profoundly disturbing.

You'd think it'd be an easy task. Feed dogs silica (dissolved, colloidal, or particulate), wait a few hours, collect urine, and analyze it for dissolved silica. Somehow, though, they managed to make a mess of it. Perhaps Earl King and his coworkers were having problems with their analytical techniques. Or maybe the analytical techniques of 85 years ago weren't up to the task. Whatever the problem was, Earl King and his coworkers decided that the answer was to get a massive amount of dissolved silica into a dog in a short period of time. Because that they ought to be able to measure.

The highly concentrated silica solution they prepared for this was alkaline, something which is not unusual for a highly concentrated silica solution to be. But, as they immediately discovered, when you send too much of an alkaline solution down a tube and into a dog's stomach, the dog vomits. A bummer for all involved.

A modern-day silica scientist (or perhaps just a more seasoned one) would have known that a silica solution, even a highly concentrated silica solution, does not need to be alkaline.[2] You can add a little bit of acid to it to neutralize it. You can even add enough acid to make the solution acidic. The silica will stay in solution, no problem.

The papers Earl King and his colleagues published don't tell us how many dogs barfed in vain before they figured this out. One gets the feeling is was a lot. But, eventually they caught on and started neutralizing their solutions. The dogs stopped puking and science was able to progress.

What Earl King and his colleagues were then able to learn was that the dissolved silica they dumped into the stomachs of the dogs was efficiently absorbed across the gut wall (in that it rapidly appeared in the urine of the dogs). But try as they might, King and his coworkers were unable to measure any real increase in the silica concentration of the dogs' blood over the twenty-four hours after pumping the silica solution into the stomachs.

How *frustrating*.

Because what Earl King and his colleagues really, really wanted to know was how quickly dissolved silica could be cleared from the blood by the kidneys. Their goal was a graph showing an initial blood silica spike followed by the long tail of a decrease. For this you need a spike in blood silica concentration. *Stupid stomachs!* they must have thought. *Can't pass silica into the blood stream faster than the kidneys can clear it.* You can imagine them wracking their brains. How could they circumvent the seeming (and now we know erroneous) problem of kidneys working faster than the stomach walls?[3]

Why, inject neutralized silica solutions directly into the bloodstream of the dogs, of course.

They did this first with bunnies (which are less expensive than dogs), with exactly the same result they then went and had with the dogs, which was that the animals almost immediately died. Too much dissolved silica in the bloodstream turned out to be a bad thing. Blood is not water. Once the highly concentrated silica

---

[2] In fact, if the solution contains a great concentration of dissolved silica, it's better if it isn't alkaline. At high pH, strongly concentrated silica solutions will be at great risk of forming silica colloids and gels.

[3] Judging from the modern-day medical literature on dissolved silica in blood (serum), this problem was an artifact of their measurement method, presumably cutting edge at the time but nonetheless not up to the task.

6.3 The Dog Days of Silica Medical Research

solution entered the bloodstream it polymerized,[4] forming flocs that blocked up veins and arteries, halting the flow of blood to such useful things as lungs, hearts, and brains. The general medical term for such an obstruction, regardless of its origin or composition, is an embolism. You don't ever want to have an embolism.

Times being more barbarous than they are today, King and his collaborators were not yet done. The next item on their agenda: intravenous injection of suspensions of particulate silica. If you happen to be reading the group's published papers, reaching this part is a definite WTF moment. As scientists, we (the authors) are all for the quest for knowledge, but there are limits and this was way over the line. All you can think is what sort of people were these who did this work and *what the hell were they thinking*.

In other words, it is hard to see why Earl King and his collaborators injected particulate silica suspensions into the veins of dogs in the wake of wreaking so much havoc with silica solutions. After their set of embolized bunnies and dogs, they couldn't for a moment have thought that injecting particulate silica would end well. (Indeed, it did not.)

Although you might guess that direct injection of suspensions of particulate silica would also induce embolisms, that isn't what ended up killing this set of dogs. Instead the particles of silica all but immediately clogged up the kidneys of the dogs, killing them all within a few hours. Most notably, despite the 300 milliliters (one and one-third cups) of liquid that had been added to their bloodstream with the injection,[5] the dogs died with bladders that did not contain a single drop of urine. That's how fast and thoroughly the kidneys had clogged.

The upshot of all this for silicosis was that if any of the particulate silica embedded in miners' lungs dissolved, it would exit the lungs via the bloodstream where it would be removed by the kidneys and excreted in the urine. Thus if miners were able to clear silica from their lungs, the researchers could monitor it. Once the miners stopped inhaling new dust, the silica content of their urine should decrease over weeks to months.

It was a nice idea, but it turned out to be impossible. When the researchers made the measurements of dissolved silica in the miners' urine, the numbers bounced around all over the place.

The researchers soon realized that silica concentrations in urine fluctuate throughout the day, spiking shortly after a person has been eating or drinking. The timing and height of the spike also differs with the amount and types of food eaten. There was no way to monitor the possible washout of silica from lungs against such a large and wildly fluctuating background.

Was all this research for nothing? Did all those poor animals die not just terribly but entirely pointlessly? It is no comfort to the animals that were involved, but the answer is no, the work did amount to more than nothing. The studies undertaken by

---

[4]Probably due to the change in pH or viscosity.

[5]This is a volume of liquid which would have been more than enough liquid to fill the bladder of dogs not the size of Great Danes.

King and his colleagues were clumsy, callous, and cruel and they failed to achieve the goals they had set out for themselves, but they did teach us things we had not previously known. Who would have guessed that we have dissolved silica in our blood almost all the time or that kidneys are excellent at clearing dissolved silica from the blood or that their being able to do so is a damned good thing? We're constantly ingesting dissolved and particulate silica in our food and drink and dissolved silica in the gastrointestinal tract crosses over into the bloodstream. If this dissolved silica was not efficiently removed by our kidneys, we'd quickly die of silica floc embolisms, just like those poor research animals nearly a century ago.

Does this excuse the experiments? For us, the answer is no. And we're still struggling to imagine exactly what kind of people were able to carry them out.

6.4 Collagen

So how did we get from thinking of silica as being bad for the body to thinking it might be necessary? The simplest irrefutable way to figure out if a substance is required by an organism is to withhold it and see if the creature fails to thrive. This is what Edith Carlisle did so successfully in her chick experiment of the early 1970s. But we still don't know why she did the experiment. Yes, because of King and his colleagues, she knew that silica could be absorbed into the bloodstream, transported about the body, and eventually excreted in the urine. But it's still a huge leap to think silica might be needed by chickens. What brought her to it?

Prior to the great chick experiment and her interest in silica, Edith Carlisle had been interested in the mineralization of bone. She spent years meticulously slicing and mounting thin sections of growing bone taken from young rats. She examined these thin sections using an electron microprobe, a nifty machine, still in widespread use even after all this time and further technological progress. With the electron microprobe, Edith Carlisle could measure the basic chemical composition of micro-sized areas of her samples. Over and over again and to her utter surprise she saw that silica was highly concentrated in bone's actively calcifying areas. To her this meant that silica was playing an active and critical role in bone growth. This, then, was what inspired her to starve some chicks of silica to see if it was true.

Go back to the photo of the chicks one month into their different diets (Fig. 6.1). The results scream loud and clear. Silica-deprived chicks can't properly biomineralize their bones. Not only do these chicks grow only short and scrawny, their skulls are deformed, and they can't do better than a bedraggled set of feathers. When you dissect the silica-starved chicks, inspecting everything very closely, you find that their bones, connective tissues, and blood vessels contain significantly less collagen than they should. As we've said, the poor things can't even hold up their heads.

Save for the bit about the feathers, the same story holds true for rats. Recently born rats raised without adequate silica also form deformed skulls and the bones, etc., they grow are also short on collagen.

6.4 Collagen

Later rat experiments went further, wounding the rats so the healing process could be observed.[6] The wounds of the silica-starved rats healed slowly and with significantly less collagen (and biochemical precursors to collagen) than the wounds of rats adequately supplied with silica.

The inescapable conclusion is that silica-deficiency makes it hard to make collagen. It's an answer that isn't yet an ultimate answer. It generates more questions than it addresses. Honestly, what on Earth does this triple helix of a protein have to do with silica? There is absolutely no obvious logical connection between the two things.

Perhaps you only know collagen as something to be injected in lips, as something to be hydrolyzed to make gelatin, or as the gelatinous gloop you get when you boil down meat scraps and bones for soup. Collagen is all this and more. Your life depends on it. Collagen is the fiber that literally holds that multicellular you together. Without collagen, you'd be a pile of goo on the floor.

Collagen is the protein that is the main component of your connective tissue, giving strength and cohesion to your skin, bones, teeth, cartilage, and blood vessels, holding your fat in place, and, not only holding your organs together, but cushioning them and, with respect to the organs in your abdominal cavity, serving as the living, pulsating tendrils they dangle on. Collagen is the material that makes tendons and ligaments strong but thin. Collagen is no less than the most important and, at about a quarter of the total, most abundant protein in your body.

Collagen accomplishes all this because it comes in thread-like fibrils that are formed by coils of individual molecules of collagen that are 300 billionths of a meter long and not quite 2 billionths of a meter wide. Chemical cross-linking between collagen fibrils means that a mass of collagen can exist as an organic matrix of considerable alignment and considerable mechanical strength.

If starving an animal of silica while it is young and growing prevents it from producing an adequate amount of collagen of sufficient quality, well then no wonder the poor thing turns out lackluster, stunted, weak, and malformed. But this still hasn't hit the heart of the matter. How is it that silica-deficient animals have trouble making collagen? Silica isn't incorporated into collagen the way, say, iron is incorporated into hemoglobin or magnesium sits at the heart of chlorophyll. Why is silica needed for the construction of collagen?

Happily, the biochemical steps involved in collagen production and assembly have long been and continue to be studied. This means that if you want to answer the question of how silica-deficiency adversely affects collagen production, you immediately have a list of enzymes that need to be present, precursors that need to be constructed, and genes that need to be expressed in order for collagen to get made. You can take a silica-starved animal (or, in these relatively more civilized times, silica-starved animal cell cultures) and methodically measure the activities of relevant enzymes, the buildup of collagen precursors, and the expression of necessary genes under different conditions to see where silica might be involved.

---

[6]Medical research in the 1970s could also be barbaric.

Not too much work of this nature has been done, but there has been enough for us to say that silica-deficient animals have lower levels of activity of prolyl hydroxylase, an enzyme also known as procollagen-proline dioxygenase. Prolyl hydroxylase produces hydroxyproline from the amino acid proline. Hydroxyproline is a major building block of collagen. If silica-deficiency impedes the production of hydroxyproline, it impedes an animal's ability to make collagen. Although this still begs the question of how lack of silica decreases the activity of prolyl hydroxylase, it brings us one step closer to the ultimate mechanism by which silica promotes collagen formation and lack of it inhibits it. Perhaps silica is, for example, the signal that the gene for prolyl hydroxylase is waiting for before it starts sending out instructions to make more of the enzyme.

It is at least clear how insufficient collagen leads to meager, malformed bones. As that major component of connective tissue, collagen serves as the fibrous organic mesh that runs through bone, reinforcing it, and both guiding and driving the formation of apatite, the crystals of calcium phosphate that make up the mineralized aspect of bone. This three-dimensional matrix of collagen needs to be exactly right in terms of orientation and richness for the resulting bone to have the correct size, shape, and denseness. When there are mistakes in the collagen matrix or its framework is too sparse, the apatite crystals that form are the wrong size, are not oriented correctly, and are chemically a bit off (having, for example, a different degree of hydration than is normal for bone apatite). The resulting bone is not only malformed, but weak and prone to fracture.

6.5 Do Human Beings Require Silica?

Although no one has ever tried to convince their university's experimental ethics committee to let them rip a newborn child from its mother's breast and raise it silica-free into bedraggledness and thus definitively settle the issue, from what we've seen in chicks and rats, the answer to whether or not silica is required by human beings should be a resounding yes. But there will always be sticklers who want human evidence of a human dietary need for silica. So there have been attempts to tackle the question in manners less direct but more humane.

Some have tried to refute the hypothesis that people require silica by testing whether we absorb silica from food and drink. If we can't, then we don't need silica. Showing that we absorb silica, in this instance, merely means demonstrating that dissolved silica passes across the lining of the gut and into the bloodstream, a much easier benchmark than determining if silica is intentionally taken up out of the bloodstream by living, breathing human cells.

For legal and other reasons, one must be more careful with human subjects than with dogs, but many such studies have been done. They have basically all shown that feeding someone particulate or dissolved silica results in dissolved silica appearing shortly thereafter in that someone's urine. These studies have shown that concentrations of dissolved silica in human blood typically average between 8 and

6.5 Do Human Beings Require Silica?

11 micromoles of silica per liter. Thus a typical person during a typical moment of a typical day will have 5 to 7 quintillion molecules of silica in each liter of their blood (that's 5×10^{18} to 7×10^{18} molecules). A typical person will also each day pee out (or micturate, if you prefer the professional term) 700 micromoles of dissolved silica, a bit more than 400 quintillion silica molecules.

Other studies have focused on the extent to which different dietary or supplemental forms of silica are absorbed in the human gut and, if they need to be dissolved, whether they tend to dissolve in the acidic regime of the stomach or in the more neutral pH environment of the intestines. Based on dissolved silica's arrival into urine or the bloodstream, most forms of dietary or supplemental silica, be they the silica phytoliths found in fruits, vegetables, and cereal grains, the magnesium silicate of some food additives, the silica dissolved in water and beer, organic-bound silicic acid, and even the colloidal silica of some supplements are all reasonable ways of getting yourself some silica. What is already dissolved stands a good chance of being absorbed in the stomach. That not yet dissolved will, over the course of the day or so it spends in the intestines, to some degree dissolve and be absorbed into the bloodstream. Thus fairly routinely throughout any given day, a reasonable amount of silica is circulating through your body (that average 5 to 7 quintillion molecules per liter of blood) and is in principle available to the cells and tissues of your body, should they choose to use it. But whether your cells have the machinery to take it up and use it is another question.

A big picture way to tackle the question of whether or not silica is necessary for human health is to ask if taking supplemental silica has any obvious positive outcomes. This is a good question because even if we need silica, it's not clear that we need *more* silica. We're already eating and drinking 10 to 50 milligrams of silicon in the form of silica every day (for men, slightly more, as they tend to be enthusiastic drinkers of beer). What possible difference could another 10 or so milligrams make?

The experiments exploring this question are the sorts carried out on reasonably large groups of people, half of whom are given supplements and half of whom are given pills identical except that they contain no silica. At the end of a month or two or three years or some other selected time frame, the researchers inspect the subjects to see if the silica-supplemented have smoother skin, thicker hair, or less brittle nails than the non-supplemented.

In one double-blind experiment,[7] researchers found that amongst women with sun-damaged skin, those who supplemented their diet for five months with 10 milligrams per day of silicon per day in the form of choline-bound silicic acid had less rough skin and less brittle hair and nails than the women of the unsupplemented control group. Another study found that women who took a moderate daily dose of colloidal silica and also dabbed colloidal silica on their faces decreased their

---

[7]Double blind means that until it is time to collate the collected data, neither the experimenter nor the subjects know who got the real pills and who got the placebos.

amount of wrinkles and increased the elasticity and thickness of their skin over the course of several months.

Thus these and similar studies that, yes, inevitably targeted women, seem to be saying that increasing your daily silica intake above normal by anywhere from 10 to 100% by swallowing a supplement can qualitatively improve the state of your skin, hair, and nails.

Studies studying cohorts of people (like nurses) who are being tracked as part of larger, longer term medical studies have noted that adults who happen to ingest more silica per day happen to have greater bone mineral density (in their hips) than those people who to ingest less. But before you head out to buy silica supplements so you can avoid osteoporosis you should know that there may be a more mundane explanation for the two seemingly related observations. People who ingest more silica per day are doing so mainly because they're eating a lot. People who eat a lot tend to be heavy for their height. Bones are not deaf to the loads they bear. People who are heavy for their height also tend to have denser bones. The relationship between silica intake and bone density may therefore be what math people call spurious, the increase in both factors being caused by extra eating rather than silica intake itself being the driving force behind greater bone densities.

The most successful approach so far for revealing the human need for silica doesn't use human guinea pigs at all, but instead grows human cells, especially cells from bone and in particular human stromal cells from bone marrow, in little dishes in the laboratory. Stromal cells, by the way, come from the connective tissue of organs. Fibroblasts, which are those cells that produce collagen (as we saw them doing disastrously in the lungs of miners), are one of the most common types of stromal cells in the body.

One recent study in fact worked with several types different cultures of human bone cells. There was a long-established laboratory culture of cells from a human bone tumor (human osteosarcoma cell line MG-63). There were bone marrow stromal cells taken from the hips of five patients, ranging in age from two to eighteen years, who were undergoing a bone marrow sampling procedure anyway for medical reasons. There were also cells from HCC1, an immortalized[8] but non-cancerous cell line of human bone marrow cells. As skin is also full of collagen and therefore potentially requires silica for its growth and general health, the researchers also cultured fibroblasts taken from human skin leftover from surgical operations.

Cultures of cells taken from such multicellular animals as humans are grown in growth media whose recipes call for salts (like potassium chloride and monosodium phosphate), amino acids, glucose, trace elements, vitamins, and other things required to keep those particular types of cells happily, healthily growing. The recipes of course also call for these things in optimized concentrations and

---

[8]Immortalized meaning that, unlike most human cells, they have the ability to infinitely propagate rather than having a set number of cell divisions allocated to them over the lifetime of the strain. Generally such unlimited cell division requires a genetic mutation of the sort that gives cancerous cells the ability to do the same.

proportions. It takes some time to figure out, but once someone has developed a good medium, it is given a name (like Dulbecco's modified Eagle's medium, or DMEM for short) and can become widely adopted. Using a widely used medium helps researchers compare the results of otherwise very different experiments. If the basic growth media used in two experiments were the same, then it was other aspects of the experiments that resulted in the differences (or failed to result in any differences) in their results.

Cell culture media don't generally include an explicit source of dissolved silica, such as sodium metasilicate (unless it is a cell culture media, such as F/2 + Si, designed specifically for growing siliceous diatom cells). So in this experiment with these different human bone cell lines, the researchers added sodium silicate to DMEM themselves and were thus able to provide as much dissolved silica as they wanted. They opted for a range of experimental treatments spanning from no-added silica to 50 micromoles per liter added silica. The outcome? The bone and skin cells supplied with silica produced more collagen than their counterparts in the no-added silica controls. The effect was strongest, at 1.8 times more collagen than the controls, in the two cell lines capable of unlimited growth (the osteosarcoma cells and the immortalized but non-cancerous bone marrow cells). The cultures of more ordinary cells produced 1.2–1.4 times more collagen than the no-silica controls. So while this does not show that collagen-producing cells absolutely required a source of dissolved silica in order to make collagen (as the no-added silica controls still produced collagen as they grew), it does show that providing these cells with silica as they grow and divide results in them producing a lot more collagen than they would without it.

A different study grew bone marrow stromal cells taken from men who were undergoing knee surgery. The cells that were grown in DMEM lacking added silica did not produce any nodules of bone mineral during the experiment. But the cells incubated in DMEM supplemented with dissolved silica did. Even more interestingly, at greater and greater concentrations of silica in the growth media, the bone nodules produced were larger and more prevalent.

During this experiment, the researchers also assayed what the cells were up to. Compared to the bone cells grown without added silica, the cells supplied with it more strongly expressed genes for alkaline phosphatase and osteocalcin, which are, respectively, an enzyme and a hormone crucial to bone biomineralization. Thus silica may have promoted the formation of mineralized nodules in these cultures by stimulating the bone marrow stromal cells to get their bone-biomineralizing machinery into gear. Thus not only does silica help stimulate the production of collagen, itself a key component of bone, silica also signals to bone-mineralizing cells that it is time to start mineralizing.

This in turn suggests that silica concentrations in our bone-biomineralizing cells are tightly regulated. That is something that would require cellular systems (like membrane-spanning transporters) specifically equipped for handling silica.

That's actually kind of charming, something to warm the cockles of the heart of every silica scientist. Even though we don't biomineralize silica, there is an extremely high chance that we have segments of our genome devoted to the careful handling of silica. And our use of silica as a signal molecule to provoke a response in collagen- and bone-producing cells in a sense is both sensible and economical. Ordinarily we'd never face a shortage of silica, even in times of famine (since water provides a reasonable supply) and, unlike heavily biologically utilized things (like calcium ions, for instance), silica hasn't already been recruited into use for other tasks within cells, tissues, or the entire body itself.

6.6 To Supplement or not to Supplement

If dissolved silica plays a critical role in collagen production, bone biomineralization, and wound healing and to boot gives us stronger hair, skin, and nails, should we all be taking silica supplements? Even as silica scientists, our level of enthusiasm for this sums to: beyond separating you from some money, it (probably) would do no harm. You'd think our position would be: *Swallow all the silica!!* But you're already ingesting amazing amounts, especially if you're drinking water, eating bread or breakfast cereal, and/or drinking beer. Although no nutritional benchmarks have been set for silica, it's unlikely that you are not just naturally getting enough. And if you aren't, why bother with something as joyless as a daily supplement when you could just drink more beer?[9]

As we have been hinting but will now show you (check out Table 6.1), your food and drink are loaded with silica. Have a gander at, for example, the silica punch that rolled oats pack: nearly 30 milligrams of silica per hundred grams of oat. And dates, yow! Almost 40 milligrams per hundred grams.

Although those are the super stars, most edible plant matter contains significant amounts of silica. Brown rice comes in at 8 milligrams of silica per hundred grams, lentils at 9, and cucumbers at 5. And because cows eat grass and grains, even dairy products contain a respectable amount of silica.

Water does, too. Fresh water, of the sort you find in rivers, lakes, aquifers, soils, reservoirs, wells, and running out of taps, contains anywhere between 50 micromoles of dissolved silica per liter and more than 800 (in other words, 3 to 50 milligrams of silica per liter). This means that water that you drink and prepare foods with is full of silica from all the silicate minerals, in soils, sands, gravel, boulders, and even entire rock formations that are slowly but constantly dissolving on the continents. This dissolved silica is fairly easy for your body to absorb across the gut wall, making water an important daily source of silica for the people who drink it.

Plants have a lot of silica because they also absorb the dissolved silica from the water they ingest and many of them use it to precipitate nanoparticles of biogenic silica known as phytoliths, two types of which are shown in Fig. 6.2. This is

---

[9]This is humor. Too much beer is bad for you. Also one of the two authors hates beer and would never recommend it to anyone.

6.6 To Supplement or not to Supplement

Table 6.1 The silica content of some common foods and beverages

Food	Milligrams of SiO$_2$ per 100 g	Food	Milligrams of SiO$_2$ per 100 g
Apples	0.4	Hummus	2.1
Arugula	6.0	Lentils, red	9.4
Avocados	1.3	Oats, rolled	27.9
Brazil nuts	0.6	Potatoes	0.9
Bread, white	3.9	Rice, short-grain, white	2.1
Bread, whole wheat	10.3	Rice, brown	8.1
Broccoli	1.7	Spaghetti noodles	1.5
Cabbage, red	0.6	Tofu	6.4
Cashews	1.3	Turnips	1.9
Cheese, hard	1.1		
Clementines	5.4	Beer	1.3–12.9
Cucumbers	5.4	Water	0.2–1.5
Dates	36.4	Wine, red	0.4–5.1

Data have been recalculated from Powell et al. (2005)

especially true of grasses (a group which includes wheat and spelt, oats, barley, rye, millet, bamboo, maize, quinoa, sorghum, and rice). Up to 10–20% of the dry weight of grasses can be silica. But phytoliths are found in other plants too, including summer and winter squashes and many other fruits and vegetables, although in less stupendous amounts (roughly 1–3% of the dry weight of these plants is silica).[10]

When you eat plants that contain phytoliths made out of amorphous, hydrated silica, you absorb some of this silica into your bloodstream. The silica has to dissolve out of the phytoliths first, however, so it is not as efficient a process as absorbing dissolved silica from water and other beverages. Not all of the phytolith silica dissolves. Instead it survives to exit your body as nanoscopic particles of glass packed in your poo, along with the rest of the non-digested or only partly digested portions of your food.[11]

This brings us to beer. Beer, especially of the more highly fermented Belgian variety, can contain appreciable amounts of dissolved silica... up to 13 milligrams of silica per hundred grams of liquid (Table 6.1). The 330 milliliters of such beer contained in a typical Belgian beer bottle would give you 43 milligrams of dissolved silica. An Imperial pint of such beer would deliver 74 milligrams of dissolved silica into your stomach. A full German liter of such a beer would deliver 130 milligrams.

---

[10]But still, that's a lot! Imagine if you were 1–3% glass.

[11]Feces contain an amazing amount of calories that you didn't absorb from the foods you eat. This makes a mockery of calorie counting. Although the people who have come up with the numbers have tried to take the inefficiency of absorption into account, it is impossible to precisely estimate and quite variable from person to person, cooking technique to cooking technique, and so on.

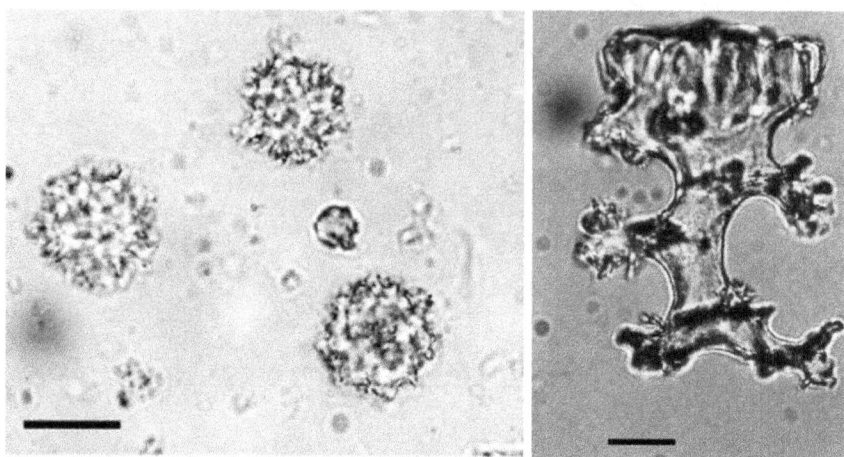

Fig. 6.2 The panel on the *left* shows a spherical phytoliths with a nodular surface from *Halopegia azurea*, a type of arrowroot from Africa. The panel on the *right* shows a considerably more specialized shape from what was likely a seed from a species of arrowroot plant (also from Africa). The scale bars in both images are 20 μm long. These image have been reprinted from Review of Palaeobotany and Palynology 107, F Runge, The opal phytolith inventory of soils in central Africa —quantities, shapes, classification, and spectra, 23–53, (1999), with permission from Elsevier

Supplements, by comparison, aim to delivery 10–50 milligrams of silica per day, often in an inefficient form, like colloidal or particulate silica, which won't entirely dissolve during their passage through your gut.

The choice is yours.

The two reasons beer has silica in it are one, even the most alcoholic of them are still more than 90% water, and two, malted barley[12] which, along with hops, is fermented to make beer, is a type of grass that takes up a lot of silica from the water in contact with its roots. To make matters ultimately more siliceous for the beer that is brewed, all this silica tends to accumulate as phytoliths in the husks of the grains of the barley. After malting of the grains, the barley is mixed with hot water for a few hours in a process, known as mashing, that dissolves a significant amount of silica out of the husks. Afterward, when the liquid and the grains are separated (and the grains washed repeatedly to collect as much of the resulting sugar as possible), not only does the dissolved silica stay with the liquid, so do any undissolved phytoliths that were released from the barley during this process. The subsequent brief boiling of this liquid (along with hops that have been added) would encourage the dissolution of many of these remaining phytoliths and any that survived that

---

[12]Malted barley is barley that has been allowed to begin germinating in order to break down some of the starches and proteins present in the grain. The barley is then blasted with heat to stop the germination process before the sprouts get too big.

would then have the weeks to months of fermentation with which to continue their dissolution.

Not all beers pack a 130 milligrams of dissolve silica per liter punch. Some come in at only 13. This tenfold range of possibility is due to a number of factors. Concentrations of silica in the water that goes into beer are different from place to place (the water in some localities has had more opportunity to accumulate silica from the weathering of silicate rocks). Some strains of barley may contain more silica than others, likewise does barley grown in soils rich in labile silica. Different brewers mash at different temperatures, for different durations, and with water of slightly different chemical properties (some being harder or softer than others, or more or less acidic). And different brewers wash the mashed grains with greater intensity than others. All of these factors plus other variations in the process of brewing and maturation of the beer add up to the great differences in the silica content of different beers.

6.7 Silica, Aluminum, and Alzheimer's Disease

The last thing we want to mention in this chapter is that there are those who believe that silica can save you from getting Alzheimer's disease. The reason? Silica can keep aluminum busy.

In contrast to silica, aluminum, which is also widely distributed substance derived from the weathering of silicate rocks, is not needed by the human body. Indeed, in fact, it is a known neurotoxin and, if it gets into your bloodstream, it gets into your brain. Although the case is far from settled, there are some signs that aluminum that has made it into the brain plays a role in the initial development or continuing progression of Alzheimer's disease.

Alzheimer's is a disease that involves the progressive degradation and death of neurons until a person's memories, humanity, and basic biological functioning slip away. Aluminum has been found in the neurofibrillary tangles of Alzheimer's victims. These tangles are microtubules that have been deformed into uselessness because one of their constituents (tau proteins) has become abnormal. By failing to function properly as a conduit for nutrients and other necessary materials within the neurons they inhabit, neurofibrillary tangles cause the neurons' deaths. Similarly, aluminosilicate particles have been found at the heart of amyloid plaques, the hard clumps of protein fragments that Alzheimer's sufferers accumulate to disastrous effect between many of their neurons.

If you've ever wondered why some deodorants (never true antiperspirants, however) loudly announce that they don't contain aluminum, well, these are the reasons why.[13] There is also concern that taking in a lot of aluminum via food and drink can lead to Alzheimer's disease. Unfortunately, for the moment, this remains a difficult issue to sort out. Because, of course, as a researcher you can't feed a group of people a lot of aluminum for decades and see if they develop Alzheimer's disease (curse those pesky medical research ethics committees). If you're trying to show cause and effect, you're stuck with largely noninvasive, observational methods.

You could, for instance, try to relate the aluminum content of drinking water and the occurrence of Alzheimer's disease. If you did this, though, you would note that populations that live in areas with high concentrations of aluminum in the drinking water don't always have a raised incidence of Alzheimer's disease. Does this mean that drinking a lot of aluminum does not increase a person's chances of tangling their neural fibrils and having globs of broken proteins congealing in their brains? Or does it mean there is another factor at play? One that could possibly defuse the danger that ingestion of aluminum represents.

Silica could be such a factor.

Silica likes to bond with aluminum. That's why so many of the silicate minerals in the rocks of the world are in fact aluminosilicates. Take for instance kaolinite and other clay minerals, feldspars (which are a major component of granitic rocks), micas like biotite and muscovite (also common components of many rocks), and zeolites. These are amongst the most common of the rock forming minerals found in the world around you and they are all aluminosilicates.

Although the chemical environment of the human body is nothing like that of typical aluminosilicate mineral forming conditions out there in the natural world, it has been shown that at high enough concentrations of dissolved aluminum and dissolved silica, aluminosilicates can indeed form there.

Aluminosilicates are not terribly soluble, so they tend to precipitate. If this occurs in the gut this is probably good news.[14] It would prevent the aluminum from crossing the gut lining, getting into the bloodstream, and potentially ending up in the brain. This may be why sometimes high concentrations of aluminum in drinking water are linked to more cases of Alzheimer's disease and sometimes not. If some of the water with high aluminum concentrations also has high concentrations of dissolved silica, then perhaps aluminosilicate formation is blocking the absorption of some of the aluminum. The silica-to-aluminum ratio of water varies from place

---

[13]All antiperspirants contain aluminum chlorohydrate or some other aluminum-containing compound that coagulates electrolytes to plug up the ducts of your sweat glands, preventing your sweat from reaching the surface of your skin. Many deodorants also contain aluminum in part to retard the growth of the bacteria whose growth generates the stink that is underarm odor. Those so-called natural deodorant crystals that you see in shops, by the way, are almost pure crystals of either potassium aluminum sulfate (also known as potassium alum) or ammonium aluminum sulfate (ammonium alum).

[14]Although in massive amounts, it would cause severe and potentially fatal constipation. In other words, there are reasons that kaolinite and other clays have been long used as antidiarrheal agents (even if the FDA no longer agrees with this).

to place because of the types of rocks that are being weathered, the acidity of the water, the extent to which clays have precipitated during the weathering process, and, again, the extent of opportunity the water has had to dissolve silicate rocks. There's plenty of room there to mess up a good correlation.

On the other hand, in the brain, the insolubility of aluminosilicates is presumably bad news. If dissolved aluminum and dissolved silica meet there and bond, they precipitate. That must be how particulate aluminosilicates can come to be found in the brain, at the heart of some amyloid plaques. Are these aluminosilicate particles a grain that catalyzes the growth of the plaques? It is hard to ascertain, but if so, then the last thing you want to happen is for dissolved aluminum and dissolved silica to meet inside your brain.

All of this leaves a person with a quandary. Ingest more silica to potentially minimize the absorption of dietary aluminum into the bloodstream (and possibly thus ultimately the brain)? Or ingest less silica so you have lower concentrations of silica in the blood and therefore perhaps less silica that will bump into aluminum in the brain?

Take this with a grain of salt,[15] but the easiest answer is to stick with silica but avoid unnecessary aluminum. Ditch the antiperspirant and avoid deodorants that contain aluminum. But unless you are a dialysis patient who has high blood concentrations of aluminum (because aluminum is hard to clear from the blood through dialysis), don't stress about dietary sources of aluminum like drinking water and aluminum-containing baking powder.

In the meantime, kick back, relax, grab a beer or some water, have some bread or eat some dates, and think about all that silica coursing through your veins, bringing your collagen to life and strengthening your bones.

Further Reading

Carlisle EM (1972) Silicon: an essential nutrient for the chick. Science 178:619–621

Han P, Wu C, Xiao Y (2013) The effect of silicate ions on proliferation, osteogenic differentiation and cell signalling pathways (WNT and SHH) of bone marrow stromal cells. Biomater Sci 1:379–392

Powell JJ, McNaughton SA, Jugdaohsingh R, Anderson SHC, Dear J, Khot F, Mowatt L, Gleason KL, Sykes M, Thompson RPH, Bolton-Smith C, Hodson MJ (2005) A provisional database for the silicon content of foods in the United Kingdom. Br J Nutr 94:804–812

Schoelynck J, Beauchard O, Jacobs S, Bal K, Barão L, Smis A, Van Bergen J, Vandevenne F, Meire P, Van der Spiet T, Cools A, Van Pelt D, Hodson MJ, Struyf E (2013) Dissolved silicon and its origin in Belgian beers—a multivariate analysis. Silicon 5:3–12

Schwarz K, Milne DB (1972) Growth-promoting effects of silicon in rats. Nature 239:333–334

---

[15]Don't take our advice too seriously. Because dammit, Jim, we're geochemists, not doctors.

Chapter 7
Of Fields, Phytoliths, and Sewage

We hate to be the ones to break it to you, but if lawns can hate, yours hates you. You mow it, probably once a week during the growing season, and then tidy up, conscientiously clearing away the clippings, bagging them up for the garbage truck. Do you know what it is you are doing? Blades of grass are stocked full of silica in the form of phytoliths, those minuscule bits of biogenic silica biomineralized by land plants and introduced briefly in the previous chapter. Every time you throw out grass clippings instead of mulching them, you're exporting silica from your lawn. Though the silica will be slowly replenished by the weathering of the minerals in the soil under the lawn, but that process can't compete against even the most mowing-averse, non-mulching lawn tender. Unless you've been replenishing it, your lawn may be by now silica-deficient. The same goes for many agricultural fields.

If you water your lawn with tap water, which inevitably contains dissolved silica, this might be enough. Rainwater isn't a good source, though. Neither is most commercially prepared "lawn food." What it really takes is spreading manure routinely, manure being recycled plants and plants being full of silica. (Bravo to you if you have had the olfactory fortitude to frequently go that route.)

As for farmers' fields, many have been farmed for centuries and most farmers do not intentionally apply silica-rich fertilizers to their fields. And so, alas, every time they harvest a crop, ta ta silica. Again, it is true that new dissolved silica enters the soil from the continual dissolution of silicate minerals within the soil. But this is also too slow a process to balance the removal of phytoliths by harvesting one to multiple times per year.

How big is this silica-stripping problem? Nearly 40% of the land area of Earth has been converted to agricultural use and the crops we grow tend to be between one and ten percent silica. That amounts to a lot of silica to go missing with each harvest. Or maybe not missing so much as passed through gastrointestinal tracts to enter the murky underworld of sewage pipes.

But then what? Frankly, we don't really know. Does this silica find its way back to field and stream? That would be a major amelioration. But if silica is removed during the sewage treatment process and trapped somewhere (like landfill) or just flushed out towards the sea, then we really are stripping nearly half of the land of one of its most vital essences.

This is a tale, then, of plants and their need for silica and the effect our farming has had on the cycling of silica over the land and down to the sea over the past 1000 to 10,000 years.

7.1 All Plants Have Silica

A factoid oft quoted in a certain subset of recent scientific papers is that all plants contain silica. Although you should never trust an absolute statement, it is true that a whole lot of land plants biomineralize silica. The most enthusiastic biomineralizers, the so-called silica accumulators, purposefully pump silica into themselves using silica transporters to achieve 1–20% of their dry weight[1] being SiO_2, mainly in the form of tiny, beautiful phytoliths such as those pictured in Fig. 6.2 in the previous chapter. But there are also land plants that are more moderately siliceous, containing no more than 1% silica by dry weight. These passive accumulators are happy to let dissolved silica simply diffuse from soil waters, across cell membranes, and into their bodies. And there are those misguided plants, like tomatoes, that actively excrete what little silica they have failed to exclude from their bodies in the first place. But for the most part, plants grown in soils or solutions containing silica will not clock in under one-tenth of one percent silica by dry weight and this silica mainly occurs as an opaline biomineral.

In other words, the world around you is jam packed full of biogenic silica.

To give you one example, assuming that it had a typical Norway spruce silica content of 0.9 percent of its dry weight and that the dry weight of the tree was 50% of its green weight, the Rockefeller Center's 80 foot tall, 10 ton Christmas tree for 2015 contained 41 kilograms of biogenic silica. That's 90 Imperial pounds of silica that were slurped up out of the ground, hoisted towards the sky, and deposited mainly within the needles of a single tree like billions of secret ornaments made of glass.

Impressive, yes, but nothing compared to the siliceous splendor of entire ecosystems. If you are wandering through a deciduous forest in Europe or in eastern North America, the 100 meter by 100 meter patch, or hectare,[2] surrounding you may contain 200 kilograms (440 pounds) of biogenic silica in live plants plus

---

[1]The dry weight of something is what it weighs after it has been thoroughly dried out in an oven at a temperature too low to cook the material (60 °C aka 140 °F is popular) or via some other similar nondestructive method of dehydration.

[2]One hectare is roughly equivalent to two and a half acres.

another 1000 kilograms (2200 pounds) of phytoliths in its soils.[3] But if tall grass prairies are more your thing, the hectare contains 70 kilograms (150 pounds) of biogenic silica in live grass and 18,000 kilograms (40,000 pounds) in the soil. Now put on rubber boots and stand in a wetland. Here the hectare holds 1000 kilograms (2200 pounds) of biogenic silica in its grasses and 16,000 kilograms (35,000 pounds) of biogenic silica in its soil and decaying vegetation.

Although there are cute microbes in the soil that are covered in silica scales (and resemble some of the vase-shaped microfossils of 800 million years ago introduced in Chap. 5), most of the biogenic silica in soils is phytoliths that have been liberated from the plants above. Annuals die. Leaves fall. Insects chew (and then poo). Cows too. There are so many ways to return phytolith silica to the soil, and not just as solid silica. Most liberated phytoliths will in time dissolve, replenishing the soil water reservoir of dissolved silica available to plants or ultimately flushed into ponds, lakes, and rivers.

7.2 Opal Phytoliths

As with other structures of biogenic silica, such as those described in Chap. 5, the production of biogenic silica (phytoliths) by land plants occurs in three basic steps. One is uptake of silica across a cell membrane and into the body of the plant. Two is transporting the silica to its site of deposition before it gels or outright precipitates. Three is using specific organic compounds to control the precipitation of the biomineral. In some cases there is also a step four, which is moving the biomineral to its final location.

Land plants have two ways of acquiring dissolved silica. All land plants, from vascular land plants that take up water through their roots, to the bryophytes (mosses, liverworts, and hornworts) that soak it in all over their bodies, have silica entering them via diffusion across cell membranes. This is just something that happens when the concentration of dissolved silica in the water in contact with the plant is higher than the concentration within the cytoplasm of the plant cells.

From there the plant can do either of three things.

It can decide that it doesn't want the silica. These plants have systems to pump dissolved silica back out of the plant. End of story.

It can decide that this slow influx of silica is enough to fulfill all its silica biomineralizing needs. These are known as the passive accumulators and the

---

[3]These soils also contain another 70,000 kilograms of amorphous silica precipitated in situ in the soil, helping to retain the silica therein for local plants to use following its dissolution under silica-depleted conditions.

amount of silica they take in is determined by physical factors such as the concentration of dissolved silica in soil water, temperature, and the permeability of the plant cell membranes to dissolved silica.

The final option that a plant has is to want more. These plants, which are known as the silica accumulators and are often but not exclusively grasses, are the ones who are using membrane-spanning silica transport proteins to pump additional dissolved silica into the plant. This is like what unicellular biomineralizers like diatoms do, but with one main difference. The silica transporters of plants directly require energy to operate. The combination of passive diffusion and active uptake of dissolved silica means that these plants acquire silica at a much faster rate than the passive accumulators (and of course than the non-accumulators).

On to step two.

Once taken into the plant at the roots, dissolved silica makes its way through the plant along with the water the plant needs to survive. In vascular plants, this means through the tube of specialized water-conducting tissue known as xylem.

To refresh your memory of high school biology, the way vascular plants manage to get water from their roots all the way to their tippy top tips (in the case of Hyperion, the coastal redwood that is the tallest tree alive today, that's 115.6 meters (379 feet) up toward the sky), isn't by pumping it. Instead, the water evaporating out through pores (known as stomata) in the leaves of a plant pulls the water up through the xylem, against the force of gravity. This rising chain of water is possible due to two of water's most peculiar properties: water's stickiness toward itself (cohesion) and toward most other materials (adhesion), for instance, the lignin that makes up xylem. These transpiration-based waterworks work marvelously well. (Did we mention Hyperion?)

But, in theory, there is a hiccup. Evaporation is distillation. It leaves the solutes behind.

As the train of water chugs through a plant and that which goes unused evaporates out the stomata at the plant's extremities, concentrations of solutes within the water of the xylem should rapidly build up within the plant. In short order, a land plant should become supersaturated with the silica that comes in either actively or passively (or both) into the plant. Their xylem should clog up with silica gel, which, if it becomes further dehydrated, would turn into solid glass, becoming entirely useless for the conveyance of water. Yet it does not happen. Land plants have learned to pump out the silica (in the case of the non-accumulators) or to use it to make phytoliths.

Step three, the controlled biomineralization of silica.

We've seen this before, in Chap. 5. In plants that produce opal phytoliths, the precipitation of silica for forming a solid occurs at specific times and places and in specific shapes, as directed, generally via chemical means, by the organism. Land plants take advantage of the concentrating power of transpiration, allowing dissolved silica to accumulate in certain enclosed spaces until it forms a gel that further condenses (meaning it loses water and forms cross-linking bonds between silica polymers) until it is a bona fide solid.

7.2 Opal Phytoliths

That there are specific organic compounds steering this process is apparent in at least two basic ways. The enclosed space within which the silica precipitates is defined by an organic membrane. The concentrations of reactive dissolved silica that build up in the xylem vastly exceed silica saturation long before the silica begins to condense. We're not entirely sure how plants manage it, but we suspect they avoid premature precipitation of silica by producing organic compounds that inhibit silica precipitation even at supersaturated concentrations.

The genes and organic compounds involved in both silica uptake and silica precipitation are not derived from the various ones used by marine biomineralizers like diatoms, sponges, and choanoflagellates. Silica biomineralization by lands plants evolved independently of silica biomineralization in protists and sponges. In fact, the transport systems that many land plants have for active uptake of silica also appear to have been evolved independently by many more than one type of plant.

Although this ability to actively take up dissolved silica is widespread throughout the evolutionary subkingdom that is plants, we did not realize it was happening until 1966. (We human beings seem to be remarkably slow on the uptake, so to speak.) That was the year that the calculations of a curious scientist revealed that barley could produce more opal phytoliths in a given period of time than it should be able if diffusion was its sole source of silica. Since then we've come to appreciate just how powerful plants can be at actively acquiring dissolved silica. We now know that wheat seedlings grown hydroponically can suck the dissolved silica in their growth medium down to concentrations that are no longer detectable. That's probably more than you could say for siliceous sponges or even some diatoms, who are ordinarily our prime candidate for world's greatest silica taker upper. We've also managed to directly observe the active uptake of silica through roots of plants like rice and cucumber.

If you want to find opal phytoliths within a plant, where is the best place to look? Short answer: almost everywhere.

In the haircap moss, to give you an example from a nonvascular land plant, long, fibrous phytoliths are threaded entirely through the plant. But this moss also precipitates silica around the walls of the cells of its leaves. If you were to gently roast away the organic tissues of the plant, an intricate, interconnected outline of every cell in the leaf would remain behind like some fantastically fine glass skeleton.

In vascular plants, individual phytoliths can be found in leaves, in xylem, in phloem (the plant's nutrient-conducting tissue), in flowers, in fruit, in seeds, in roots, and in the stiff, hair-like trichomes that protrude from many outer surfaces. Some of these phytoliths have been produced inside of cells and sometimes they have been produced in between them. The shape of the phytolith differs accordingly. The ones produced in between cells tend toward thin and spindly, the spaces in between cells being relatively long and narrow. But the phytoliths formed inside of cells tend more toward sphericality. Some phytoliths even take on the form of flat glass plates.

This wide array of possible morphologies and the details that go into them mean that plant phytoliths can be really distinctive. Each type of plant that produces phytoliths produces such reproducibly specific ones that archaeologists can peruse through otherwise nondescript remnants of plants foraged, grown, harvested, stored, prepared, and consumed or otherwise used ages ago and tell rice from squash from wheat from cotton. Phytoliths are also useful when you're in that awkward situation of having some Iron Age dung and you want to know if it came from cattle or sheep. (The diets of cattle and sheep are not entirely the same and phytoliths work just fine for distinguishing the two.) Additionally, phytoliths found in old soils or in lake sediments are helpful for reconstructing what sort of vegetation an area used to have back in its wild old days or at a time when the local climate was somewhat different than it is today.

7.3 The Benefits of Opal Phytoliths and of Dissolved Silica

Although it is amusing to think otherwise, the raison d'être of phytoliths is not to help out the archaeologists. They exist because it is more beneficial to most land plants to produce them than to not. Originally, we assumed that plants produced silica phytoliths simply to avoid death by autosilicification. Now we know that plants can actively exclude silica from their bodies, so this can't be the benefit. And who knows, maybe, like animals, land plants have found biochemical uses for the dissolved silica flowing through them. Maybe at this point, silica has wormed its way into being an essential nutrient for land plants.

They physical benefits of phytoliths are easy to imagine. Rigidly solid silica phytoliths could be great as structural support. That's something any land plant could appreciate. All but the laziest land plants stand up against gravity, skirmish with wind and rain and maybe snow, and plow through ground that can be as hard as concrete. Why would they waste time and energy excreting silica back out through their roots when they could place glass bricks here and there within their bodies as reinforcement? As an added bonus, they might not have to make as much of metabolically more expensive molecules like cellulose and lignin, the organic polymers that make plants tough, woody, and strong.

Structural integrity isn't just about standing tall. It can also be about remaining broad and open. The stiffness opal phytoliths impart to leaves, for instance, helps them remain unfurled even when they become slightly dehydrated on a hot, dry day. Such avoiding rolling themselves up like a carpet prevents them from depriving their chlorophyll of light. This is a win, not just for the plant but for all of us. The whole point of a plant is to photosynthesize as much as it can, bringing energy and organic matter into the biosphere for all of us to use.

Opal phytoliths, being the abrasive shards of glass that they are, also discourage grazing. Although their main target is small, nibbling grazers like insects, phytoliths are part of the reason that an elephant goes through six sets of teeth over the course of its life and part of the reason that grass chewers and grain munchers like horses and rodents have teeth that never stop growing in.

Even to those of lifelong tooth growth, phytoliths are a challenge they'd prefer to avoid. Give them a choice and rabbits and locusts will eat grass grown without

7.3 The Benefits of Opal Phytoliths and of Dissolved Silica

silica instead of grass grown with an ample supply. Not only is this glass-less grass less crunchy and therefore less work to eat, less of it needs to be eaten by whatever is eating it, be it bird, mammal, reptile, or insect. Any grass that isn't 10–20% glass is closer to being 100% food. Eating glassy grass means more time, energy, and tooth enamel expended on chewing and digesting, leaving less time and energy for growth, reproduction, and other frolics. From the plant's perspective, this is a major score. The more it embeds itself with phytoliths, the more it depresses the rates of growth and reproduction of the populations of the creatures that eat it.

Silica also protects plants from attack by fungi and disease-causing bacteria, in ways that show that not only is particulate silica a good thing for a plant to have inside of itself, so is dissolved silica. For one, phytoliths deposited just under the plant's outer cuticle act as a physical barrier to infestation. But plants that contain a lot of dissolved silica are also better at producing phenolics and other antibiotic, antiviral, and antifungal compounds in response to infection. This is because dissolved silica is being used as a signal that triggers the expression of the genes involved in the construction of such defensive metabolites. Dissolved silica also enhances the chemical activity of enzymes, like peroxidases and chitinases, that plants also deploy in their battles against invading pathogens.

A sufficiency of silica also helps plants deal with abiotic stressors like metals, salt, heat, freezing temperatures, drought, lack of nutrients, and ultraviolet radiation. Some of the work that silica does in these cases is strictly chemical in the most basic sense of the term. Dissolved silica present in a plant's roots can bind to and co-precipitate metals and certain salts that come into the plant via diffusion or other mechanisms. Thus the water in the roots is scrubbed of some potentially hazardous materials, preventing them from incurring further up into the body of the plant and accumulating to the point of toxicity.

Silica phytoliths help in arid conditions by slowing down the rate at which water travels into, up, and out through a plant via transpiration. This is partly because structural support from phytoliths allows leaves to grow thicker, decreasing the ratio of the area of each leaf's surface to its volume, thereby decreasing each leaf's tendency to lose water through evaporation relative to its ability to store it. Not only does this diminishment in transpiration minimize water loss and reduce the rate at which plants dry out the ground they are growing in, it minimizes the uptake of waterborne salt into the plant. Less salt taken in means less work excreting excess. In saltier soils, such as in salt marshes, this can mean life or death.

A further boon of silica to plants in salty places is that dissolved silica increases the activity of their antioxidant enzymes, and this decreases the permeability of their cell membranes. This means a plant can sit in somewhat salty water without having all of its own water sucked out of it. So here again, dissolved silica is providing an unexpected but potentially life-saving service to the plant that has it.

7.4 Is Silica an Essential Plant Nutrient?

Given all that, do plants require silica in order to survive? Is it an essential plant nutrient like nitrogen or phosphorus?

Here a debate rages. Those not on silica's side point out that you can grow even the most enormously siliceous land plant (such as horsetails)[4] silica-free just fine… at least within the fairly sterile, stressor-free confines of a laboratory. Ergo, silica is not an essential nutrient for plants. Clearly, it is not required for the core metabolic reactions you can strip a plant down to and have it still grow and reproduce. But starve a plant of nitrogen, phosphorus, requisite trace metals, and necessary vitamins, for instance, and it will die, no question about it.

But Team Silica would say that dissolved and particulate silica, at minimum in the ways described in the preceding text, play key roles in plant biology and physiology. Plants may not automatically die when deprived of silica but there is no question that things for plants go suboptimally without silica out in the cold, cruel, challenging world, sometimes fatally so. To claim that silica is not an essential nutrient for land plants because you can grow plants without it if you protect them from adversity is to define the term essential nutrient with a certain narrow-minded, short-sighted pigheadedness.

Our vote, in case you are wondering, isn't going with the pigheads. We're heading out right now to buy proper manure for our lawn.

7.5 Impact of Agriculture on the Silica Cycle

That is why plants need silica, how they acquire it, and what they use it for. Now to get back to how the human race has been inadvertently stripping easily biologically available silica out of much of the land.

Some plants live a long time, but most vegetation doesn't. If it hasn't been chewed off and eaten, it dies off in autumn, in these cases, rapidly returning plant matter (including phytoliths) to the soil. Most of the silica phytoliths dropped into the soil from plants dissolve within a few years, returning dissolved silica to soil water where it can be taken up by other plants, used up in the formation of secondary silicate minerals like clays, or flushed out into rivers that make their way to the sea. In the natural world, over the long term, the reactive silica that leaves a plot of land either by being flushed out or by being trapped in a mineral not easily redissolved is more or less replaced by silica that is dissolved from primary silicate minerals during weathering. But that is a slow process. So most of the production of

---

[4]Also known colloquially as scouring rushes (for they are so siliceous you can scrub pots with them) and, to science, as *Equisetum*.

7.5 Impact of Agriculture on the Silica Cycle

phytoliths in a plot of land is supported by silica that has been hanging around for a while and has been taken up by plants and made into a phytolith and then dissolved again already possibly multiple times.

Farming disrupts the looping of silica between soils and plants and back again because instead of letting phytoliths return to the soil at the end of the season, we haul them off to the supermarket, ultimately to flush them down the toilet. To make matters worse, the crops we grow most prolifically tend to be amongst the most prolifically prolithic. That is to say, they tend to be silica-rich grasses like cereal crops and sugar cane. It would be hard to devise a better means of removing silica from these soils, short of removing the soil itself.

Here are two examples of the numbers. Cultivated rice takes up 1100 kilograms (2425 pounds) of silica per hectare per year. Sugar cane removes 640 kilograms (1410 pounds) of silica per hectare per year. Most of that goes with the crop when it is harvested.

Natural ecosystems are nowhere near so silica-sucking. Even those most siliceous, like grasslands and tropical forests growing on highly weatherable silicates, take up generally much less than 150 kilograms (330 pounds) of silica per hectare per year. And, as they tend to provide lunch to herbivores on the spot, they are kind enough to give an equivalent amount of silica right back to the soil.

When you crunch the numbers, what you find is that the harvesting of crops removes roughly 48 billion kilograms (106 billion pounds) of silica from cropland each year. That is stupendous as a standalone number and it is stupendous in the grand scheme of things. It's comparable to the amount of dissolved silica that the rivers of the world, taken together, deliver to the oceans each year. Likewise, it is similar to the net[5] amount of dissolved silica liberated from silicate minerals each year through chemical weathering.[6] Returning straw to fields to rot ameliorates some of this loss, but it's a minor gesture in the face of the massive, intensive, industrial-scale agriculture we carry out, capable of feeding our current 7.4 billion souls plus possibly a couple of billion more, not to mention 50 billion chickens, turkeys, ducks, geese, and other poultry, 1.5 billion cattle, and 1 billion each of sheep and pigs.

This is a fact almost as new as it is staggering. What we're doing, it's unprecedented on this Earth. Since we human beings do not have much in the way of long sight, it might have become our business-as-usual, but it is certainly not the planet's.

---

[5]Net because something like two-thirds of the gross amount of silica weathered out of rocks becomes incorporated more or less permanently into clay minerals.

[6]Physical weathering, in contrast, refers to only the mechanical breakup of rock material.

7.6 The Growing Creep of Silica Removal

It wasn't always like that. Our activities did not always massively strip silica out of soils and ecosystems on either regional or global scales. In part this was because there just weren't so darned many of us.

Before we learned to cultivate plants and livestock, we hunted and gathered. Hunting and gathering requires a lot of acreage per person because that acreage requires a time at rest after a group of humans has gnawed its way through. The ecosystem's plants and animals need to regrow, reproduce, and otherwise recover before the next onslaught, otherwise the system will run down and everyone starves. It also takes a lot of natural ecosystem to produce enough food edible to humans to feed even a small group of people. We don't subsist so well on wood or on most leaves. Thus, as amazing as it seems, even in its most primeval state, the planet Earth could not comfortably sustain even *10 million* hunting and gathering human beings. In fact, the real number is *likely far less*.

Take a moment to imagine that: being one of fewer than 10 million people on this Earth. You can hear Shanghai, Karachi, Beijing, Lagos, Delhi, Tianjin, Istanbul, Tokyo, Guangzhou, Mumbai, Moscow, São Paolo, Shenzen, Jakarta, Seoul, Kinshasa, Cairo, Lahore, and the greater metropolitan areas of Mexico City, New York City, Los Angeles, and Bangkok, among others, laughing. Every single one of those places contains more residents than that all on its own. How wild, beautiful, and quiet the Earth must have been for most of our species' 200,000 year tenure. Forget about 10 million individuals, in fact. Most estimates hold populations below three to four million people until about 10,000 years ago.

So a size somewhere between 1 and 10 million people is roughly where *Homo sapiens* was 5000 years before that, a date designated more scientifically as 15,000 years Before Present (BP) or 13,000 Before Common Era (BCE). If you wanted to pick a moment when things first started to go south for the silica cycle, you could do worse than pick this one.

Environmentally speaking, 15,000 years BP was a time of change. The last ice age's peak had been passed a few thousand years before and deglaciation was in full swing. Across the Earth, temperatures were rising. The northern hemisphere ice sheet was retreating to fully reveal Canada, Scandinavia, Great Britain, and Ireland and the Antarctic ice sheet was losing some serious height. Mountain glaciers all over the world were shrinking. Along with all that ice melting, came the rise of the seas far enough and fast enough onto land to be shocking within the normal span of a human lifetime (hence, potentially, the preponderance of flood myths in various human cultures). At the same time, remarkable shifts in climate, vegetation patterns, and ecosystems were underway all over the globe.

This was also when, roughly speaking, people had the bright idea, for the first time in human existence, that if they planted seeds of things they liked to eat, they would know exactly where to return to a few months hence to find them growing in abundance. Was it genius, that endeavor that is at least two percent inspiration? Or

7.6 The Growing Creep of Silica Removal

was it more like laziness (How can I get that work done with less personal effort?) tinged with mere curiosity (Will it work?)?

The dawn of agriculture was likely all that plus just an idea whose time had come. The human brain, by then equipped with sufficient reasoning power, had finally accumulated enough culture (meaning knowledge, understanding, methodologies, philosophies, habits, and experience developed over generations) to not just think of purposefully planting things but to figure out how to make it work. The end result of the initial successful agricultural experiments (perhaps just a handful of seeds of a useful food tossed into a known location) would have been a bit of high-density food gathering that required less time and walking around while cold, wet, tired, and hungry to accomplish.

It's doubtful that the human race knew it at the time they first started planting things, but they'd just broken through the barrier that is the Earth's natural carrying capacity for human beings. It would now take less land to feed more people (at the expense of expanses of wilderness and the natural, unperturbed cycling of silica). Of course there were bumps and hiccups (like famine, malnutrition, serfdom, and slavery) along the way (agriculture is no easy thing to invent from scratch), but as we got better at growing more food reliably, the number of people that could be fed began to increase and the human race began to grow. Slowly, at first. But then, starting about a thousand years ago, like a rocket.

The initial inspired/accidental/lazy/inevitable invention of agriculture was also not an isolated stroke of genius (and disruptor of the silica cycle) that spread out from one point of origin as people wandered. Agriculture started popping up all over the place somewhat simultaneously, again, as if human consciousness and culture had finally gotten to the point where they were capable of dreaming it up and dream it up they did almost simultaneously all over the place. Over the course of a few thousand years and at up to 30 different independent locations stretching from the Fertile Crescent,[7] to China, to Southeast Asia, to South America, Central America, and Eastern North America, and to the savannahs and highlands of Africa, the initially haphazard cultivation of wild plants by nomadic people transformed itself into the high-intensity cultivation of domesticated crops and husbandry of domesticated animals by people tightly tied to land they had also entirely domesticated.

Excavations of archaeological horizons in the Fertile Crescent have revealed that already by 12,000 years ago, cultivation of goat grass and several other wild progenitors of domesticated wheat, wild lentils, wild barley, and wild grass pea had been going on for some time. If you follow the archaeological record up in that area over the few hundred years that followed, you'll see increasingly more familiar (that is, more domesticated) versions of crops beginning to appear. Meanwhile, over in Asia, rice was domesticated by 10,500 years ago, meaning that cultivation and

---

[7]A sickle-shaped swath of land encompassing the floodplains of the Nile in upper Egypt, the Levant along the Mediterranean Sea, and the land around the Tigris and Euphrates down to the shores of the Persian Gulf.

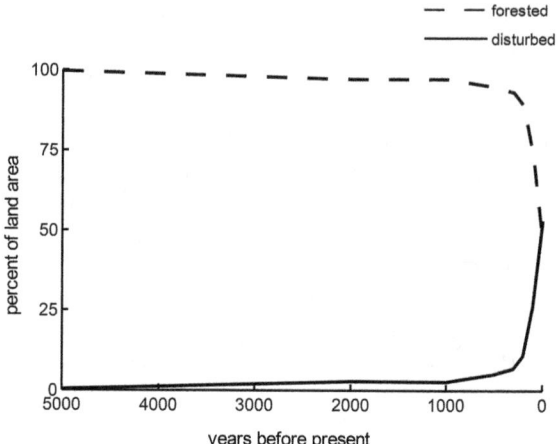

Fig. 7.1 Pace of the change in land use in temperate zones from 100% forested 15,000 years ago to the heavily cultivated (ecologically disturbed) of today. The data plotted are from Goldewijk et al. (2011) The HYDE 3.1 spatially explicit database of human-induced global land-use change over the past 12,000 years. Global Ecol Biogeogr 20:73–86

the process of domestication had begun some centuries before. Over in the New World, squash was busy being domesticated independently at sites in Mexico and in South America between 10,000 and 9000 years ago, again suggesting that these people had started growing things some time before.

This domestication of crops and the associated domestication of cows, pigs, goats, sheep, water buffalo, yaks, alpacas, honey bees, chickens, turkeys, geese, horses, donkeys, cats, and so forth has been a pretty amazing accomplishment, and is one that is still ongoing. But the truly staggering feat has been the domestication of the land.

As it was in lockstep with the size of our population and with the intensity of our interest in agriculture, the domestication of the land started slowly, almost gently. On Fig. 7.1 you can see that the majority of Earth's land was forested until only recently; there's only so much damage to the planet that a couple of million stone aged people can do. But the gentle pace picked up roughly 1000 years ago, when humanity broke through some key barrier in terms of population size, knowledge, and technology. Then, about 500 years ago, the slashing, burning, draining, digging up, and plowing under exploded truly exponentially in intensity. These days, the growing of crops and the tending of animals has been firmly established on every single continent except for Antarctica (and, given global warming, you can be sure that we've got this continent in our cross-hairs), to the tune to of our having converted 40% of Earth's land area from wilderness to field and pasture. (And you can believe that most of the remaining 60%, like all of Antarctica, a lot of Canada, most of the Saharan desert, is too extreme an environment (too steep, too dry, and/or at too high a latitude) to be hospitable to agriculture.)

7.6 The Growing Creep of Silica Removal

Can you imagine how much slashing and burning, felling of trees, digging up of grasslands, and draining of swamp and moor[8] that it took to convert 40% of the land to field and pasture?

Although we politely refer to it as a *change in land use*, it's nightmarish, our disruption, often to the point of destruction, of the ecology and biogeochemistry of nearly half the Earth's surface in order to grow food for ourselves. Never mind the ongoing mass extinction of animals (including fish and insects) and plants, through our agricultural activities alone, we've significantly altered the rates of the weathering of silicate minerals in soils, the concentrations of dissolved silica in soil waters (and, as the next chapter will explain, lakes and rivers), the standing stocks of biogenic silica built up in the soil reservoir, and the amount of silica transported to the ocean to support the silica cycle therein.

We silica scientists do our best to study it. We even (foolishly) tend to think of it as being normal. But the cycling of silica from rocks through soils, soil water, plants, lakes, and rivers through to the ocean we can observe today is not pristine. Like all other major biogeochemical cycles within which all life on Earth exists, because our domestication of the land, the silica cycle is seriously out of whack.

7.7 Let's Go for a Walk Through Time

One way to look at this out-of-whack-ness is to trace how the silica content of a patch of land would have changed over the centuries as it went from being wilderness to being an intensively farmed field. In fact we can even start with land in a sort-of pre-wilderness state—a forest in Europe that sprouted out of the barren ground exposed by the retreat of the northern hemisphere ice sheet at the end of the last ice age as it melted back until it was no more than the meager pile upon Greenland.

In the beginning, all would have been glacial debris: the rock flour from glacially demolished rocks, sand, gravel, cobbles, and larger rocks. But soon, in addition to plants that we might call fast-growing weeds, oaks, beeches, birches, spruces, pines, willows, and other trees would have begun poking their noses out of the newly liberated ground. Thank bird droppings, if you will, and other perhaps most romantic forms of seed dispersal. Over the course of centuries, that ground would have grown thick with more and increasingly bigger trees and would have come to host a lush understory of ferns, mosses, grasses, and so many other plants. With each twig added and every leaf unfurled, a biomass of tons of biogenic silica phytoliths per hectare would have started to accumulate where for tens of thousands of years before there had been only barren ground buried under miles of ice.

The first consequence of this fervid phytolith production (and accumulation) by the growing forest would have been the net extraction of silica from the waters in

---

[8]Europe alone must have millions of miles of drainage ditches drying out its land.

the soil. As that new forest grew up toward maturity, its soil waters would be depleted in dissolved silica because the forest would be removing more silica than it would be recycling back to the soils. Soil waters would have been thus somewhat depleted in dissolved silica and would have little to export to streams and rivers on their way to the ocean.

But after several hundred years, the forest would have been done maturing. Growth and death would have come into balance and likewise so would have the creation and recycling of phytoliths. The forest would still have been producing plenty of new material, but it would have been producing only enough new material to make up for its losses.

Because the total amount of phytoliths existing in the forest would have stopped increasing, the soils would have started exporting dissolved silica to rivers exactly as fast as dissolved silica was being added to them via the weathering of silicate minerals. That's just the requirement of mass balance (what comes in, must go out unless the forest and the amount of phytolith silica stored in its soils is changing in size).

When everything was at steady state in this forest, it would have held per hectare nearly 1000 kilograms of dissolved silica in soil water per hectare and 70,000 kilograms of amorphous silica (including phytoliths from plants, siliceous soil microbes, and silica that precipitated without biological intervention). In addition, it would be exporting as much silica toward the ocean as it was receiving due to the weathering of silicate minerals in its catchment area.

Then along came some agrarians to grow crops and to pasture animals. They would have likely done as they still do when starting in a new place: burn the forest down. As the ashes from all those plants cooled down and disintegrated, an entire forest's worth of phytoliths would have spilled out, entered the soil, and begun dissolving. This would have been a huge, one-time input of silica to the soil and, initially, there would have been no plants left in the ground to take up that liberated silica.

A rapid increase in the dissolved silica concentration of soil water would have resulted, followed by a time of elevated export of dissolved silica into nearby lakes and streams via rainwater flushing through the soil. But this extra silica in the soils would have only been temporary, lasting only as long as it would have taken for the dead forest's phytoliths to dissolve (50–100 years at best).

When the agrarians started planting, the first crops would have started to remove some of this extra added silica. As agriculture got a late start up in these more northern climes, somewhat after the start of the domestication of grains, many of the first crops that would have been grown on this previously forested, post-ice sheet land would have been grains like barley, emmer wheat, and einkorn wheat and crops like flax. These particular crop plants happen all to be silica accumulators considerably more adept than forest at extracting silica from the soils. Thus this first farming would have sent the silica content of the previously forested soils into a decline that could have been made much worse by harvesting, unless those fields were being fertilized with the manure of those humans and other animals that were eating the crops.

The field at that point would have already been no longer a fully functioning ecosystem (although things weren't yet as bad as they are today in the days of herbicides, pesticides, steel, tractors, and combine harvesters). Harvesting would have robbed the soil not only of their source of phytoliths, but of the plant litter that sustained the insects, animals, and microbes that normally live in soil. And if all those critters weren't eating up the plant litter and then breathing their carbon dioxide into the soil, soil waters would have been correspondingly less acidified. Rates of silicate weathering would have slowed down. This, in turn, would have meant less of a supply of new dissolved silica for the plants growing in those soils.

Over time, less input of new dissolved silica, and less old silica being recycled out of phytoliths would have driven the soil waters of the field to become depleted in dissolved silica. As an area of land, the field would have begun to deliver less dissolved silica to nearby rivers and streams and the crops growing in the field could have become silica-limited enough to become susceptible to drought, grazers, and disease. Crop yields would have started to drop. At this point, concentrations of dissolved silica in the soil would have been 250 kilograms per hectare (a quarter of its previous value) and the amount of amorphous silica in the soil would be as low as 30,000 kilograms per hectare (half of what it had been before). And that's pretty much where it would be at today.

If the deforested land had been set aside for pasturing animals instead of growing crops, the numbers would be at best about 10% higher. Although you can argue livestock tend to drop their droppings where they are pastured, in practice silica is often distilled out of pastures in the form of hay harvested and used as fodder elsewhere. Likewise, and almost hilariously, muck has always been carted around from place to place and today it is even traded internationally. Some fields and pastures might win in this game, but the rest are having their silica and other nutrients distilled away.

It's hard to make a good estimate at this point because not enough studies have been done, but if this story can be applied to the 40% of Earth's land area that has been converted from wilderness to field and pasture, the amount of biogenic and other types of amorphous silica in soils has decreased by one-fifth (20%). That is a lot and the upshot of it is that there is a lot less recyclable silica now in soils to serve as a buffer to replenish the pool of silica easily available to plants. The decreased weathering rates in agricultural soils further pushes the plants of terrestrial biosphere towards silica limitation. That's a lot of land where plants, because they may be lacking in silica, are especially susceptible to dessication, infestation, and disease.

The other problem with the agriculturally induced changes in the silica cycle on land is that, due to the lower weathering rates, notably less silica is being delivered to rivers, lakes, estuaries, and ultimately the ocean, much to the chagrin of siliceous plankton like diatoms which depend on it in order to exist, of all the zooplankton and fish that depend on diatoms as food, and of all the bigger fish, whales, dolphins, seals, and fishermen (and fisherwomen) who depend on the diatom-eating zooplankton and fish.

7.8 Silica in Sewage

So where does all the silica go?

We're willing to wager that you've never pondered the fate of the dissolved and particulate silica you flush down the toilet.

Meaning, the short answer is that silica from plants harvested then eaten goes to a sewage treatment plant.

But then what? Does the dissolved and particulate silica that has passed through you effectively disappear from the world because it has become sequestered into some strange stockpile of treated sewage? Or does it gush back out into the light of day in effluent effluxing from a pipe? We may not be silica biomineralizers, but we'd like to know where an ocean input's worth of silica is ending up each year. We're stakeholders in this too. The optimal growth of our crops may be riding on this and we have billions of mouths to feed.

Shall we go through it step by step?

Although other methods of sewage treatment exist (such as letting sewage seep, below ground, through wetlands specially constructed to break it down), modern sewage treatment is generally parsed into steps, primary to tertiary. These steps separate liquids, fats, and solids, decompose organic matter into carbon dioxide and inorganic nutrients, contain toxins, neutralize pathogens, and strip out excess amounts of inorganic nutrients before the effluent is released out into the environment. What happens to dissolved silica and solid silica in phytoliths during each of these steps is a puzzle that, understandably, has yet to be intensively investigated, but is still worth briefly pondering here.

The primary treatment that sewage is subjected to is time spent slowly and non-turbulently flowing through a receptacle where solids settle and fats, oils, and grease rise to the surface. The fats, skimmed off and collected, often get turned into soap.[9] The liquid portion of the sewage passes on to the secondary treatment phase. The sludge that has settled to the bottom does not.

While dissolved silica in the sewage thus gets through to phase two, silica phytoliths are more likely to be scavenged into sinking aggregates of goo[10] and to therefore join the sludge, which is diverted into special tanks and allowed to rot anaerobically for six to nine months. There has been a lack of volunteers to study the fate of such phytoliths. If temperatures are warm and dissolved silica concentrations remain unsaturated, they probably dissolve. However, if the phytoliths are thickly coated with sludge, they probably don't.

Any phytoliths that are in the sludge and don't dissolve hit a dead end in the silica cycle. At the end of the six to nine months, the liquid is siphoned off and any

---

[9]We're not kidding.

[10]Phytoliths are too small to sink rapidly on their own. But clumped up with the goo that becomes sewage sludge, they'll hit bottom of the settling tanks within an hour or two and that's all it takes to prevent them from passing on to the secondary stage of sewage treatment.

7.8 Silica in Sewage

solids still remaining are deemed undigestible and presumed toxic. They're dried, sealed in a leak-proof contained, and buried in a landfill.

Interestingly, though, as the sludge was fermenting, biogas bubbled out and may have been collected for use. This mixture of the byproducts of the anaerobic digestion of the organic matter in the sludge contains methane, carbon dioxide, hydrogen sulfide, and trace amounts of siloxanes.

Siloxanes (because, honestly, who knows such things?) are organosilicon compounds containing oxygen and saturated with hydrogen. Exclamation point! All silica scientists' ears have just perked up. Where on Earth did they come from? How have they gotten into biogas produced from rotting sewage?

It is so tempting to leap to the conclusion that siloxanes were produced from silica dissolved from phytoliths during fermentation. But in the real world, the siloxanes found in sewage biogas come from silicones such as dimethicone, aka polydimethylsiloxane, found in shampoos, conditioners, moisturizers, and cosmetics. Silicones also find their way into sewage because they are added to processed foods and oils as antifoaming agents and anti-splattering agents. Silicones also come from laundry detergent, and may also be laundered out of water-resistant fabrics or washed off brushes used to apply silicone-containing sealant.

Siloxanes are fancy molecules. The ones in sewage biogas range in from the simple trimethylsilanol ($C_3H_{10}OSi$) to the hexagonal dodecamethylcyclohexasiloxane ($C_{12}H_{36}O_6Si_6$). The two most common siloxanes produced are octamethylcyclotetrasiloxane ($C_8H_{24}O_4Si_4$) and decamethylcyclopentasiloxane ($C_{10}H_{30}O_5Si_5$), also known as D4 and D5, presumably for their number of silicon atoms. When these siloxanes oxidize, such as in the combustion chamber of a biogas powered engine, carbon dioxide, water, and particles of silica result.[11] Some of the silica deposits itself inside the combustion chamber but most of it is expelled to the atmosphere along with the rest of the exhaust. Provided that they don't get inhaled or ingested by human and nonhuman animals, these silica particles that are somewhere between 5 and 100 billionths of a meter in size (5–100 nanometers) dissolve in rainwater, in soils, or in the waters of lakes, rivers, or the ocean and the silica takes up an active place in the natural biogeochemical cycle of silica.

The silica nanoparticles that are swallowed either dissolve in or pass through the gastrointestinal tract á la silica phytoliths. Those that are inhaled may, like dust-sized particles of silicate rock or, worse, like fibrous ones like asbestos, damage the lungs by triggering the woeful immune cascade that can lead to silicosis (as described in Chap. 6). Before that, they may give you asthma or just generally cause inflammation in your lungs. They may also cause cancer by mutating DNA in the first place and in the second place by driving the expression of genes that support the growth or spread of cancerous tumors. So don't inhale deeply and frequently of biogas tailpipe emissions, which is more than sound advice for any sort of tailpipe emission.

---

[11] Here are the equations describing the combustion of D4
$$C_8Si_4H_{24}O_4 + 16O_2 \rightarrow 4SiO_2 + 8CO_2 + 12H_2O$$
and D5,
$$C_{10}Si_5H_{30}O_5 + 20O_2 \rightarrow 5SiO_2 + 10CO_2 + 15H_2O.$$

All of that was just the first stage of sewage treatment. There are still two more.

The secondary treatment of sewage consists of letting microbes have at the liquid portion of the sewage that was separated from the sludge during the primary treatment. Bacteria, fungi, and protozoa in aerated tanks chow down on dissolved and colloidal organic matter (such as but not limited to urea), converting them into carbon dioxide and inorganic nutrients like nitrate, nitrite, ammonia, and phosphate. During this stage, if there are any phytoliths that remain in suspension in the water, it is likely that they dissolve.

A tertiary treatment may follow. Although tertiary sewage treatment can consist of chemical removal of the excessive nutrients present in effluent, oftentimes biology does the job. In this case, the effluent is percolated through lagoons or ponds with thriving communities of algae (microscopic and macroscopic), aquatic plants, and invertebrates that remove most of the nutrients before the water is released out into the real world. This tertiary treatment takes time and a lot of space and is most needed in places where there is considerable pressure to use the land for housing instead, so it is far from always undertaken. Which is a shame, for excess nutrients released to waterways tend to clog them by fueling the blooming of noxious (and often noxious smelling) plankton (as to be discussed in the next chapter). It would be better to clog up (and stink up) tertiary treatment ponds.

If there is dissolved silica in the tertiary treatment ponds, diatoms will likely bloom in them, thereby removing the dissolved silica from the water to be discharged into the environment. This is also a shame. Dissolved silica isn't like ammonia and phosphate. In excess, it doesn't generally drive harmful algal blooms. And in fact having enough dissolved silica in your waterway can actually prevent excessive amounts of ammonia and phosphate from spurring noxious plankton. The diatoms, happily growing with their silica, are able to out compete the bad guys for the excess nutrients. But only when there is enough silica for diatoms to grow.

Anyway, any silica buried in the sediments of these tertiary treatment ponds (in the form of diatom frustules) is either delayed from its return to the natural silica cycle or prevented entirely from its return. A mixture of both things likely occurs.

At the same time, silica in wastewater can be a serious concern. Not the stuff that comes in naturally, but excessive concentrations of dissolved, colloidal, and particulate silica related to the semiconductor industry. Polishing up silicon wafers results in the release of used polishing slurries containing impressive amounts of silica. In addition, manufactured silica nanoparticles are all the rage these days. You'll find them in foods, in toothpaste, in fabrics, and as deliverers of drugs, all places where they are likely to find their way into municipal wastewater and then perhaps the environment where they are of insidiously small enough size to potentially do harm (such as with those biogas tailpipe emissions of silica nanoparticles).

And so a lot of effort has been put into devising means of removing silica particles from industrial and municipal wastewater streams via filtration and via chemical flocculation. Any phytoliths present in these waste streams would also be captured and removed, blocking the natural silica's return to the environment.

As you can see, silica in sewage has developed its very own dynamic biogeochemical cycle, worthy of serious study.

7.9 A Plea for Hardy Souls

And so this chapter ends on a disturbingly unbalanced note. We know that our agricultural practices are depleting the soil of nearly half the Earth's surface of its silica. Partly to blame are the lower rates of silicate weathering in the soils of farmed fields than natural ecosystems. But a huge amount of silica, mind-boggling even at the global scale, is being carted away in harvests and we don't have a good understanding where this silica is ending up.

This is a concern. We need to figure this out. Plants, and especially the typically silica accumulating plants we grow as crops, need silica to be strong and healthy. If only (and certainly not really only) for this reason, we need silica to be present in abundance in the soils.

So here is our plea: pick up the torch. Become a scientist. Become a *silica scientist*. For those of you especially who have had the olfactory fortitude to frequently fertilize your lawn with genuine manure, there is so much work to do to illuminate the fate of phytoliths in crops and food. And by this we mean, track the fate of silica through the sewage systems of the world.

Further Reading

Carey JC, Fulweiler RS (2015) Human appropriation of biogenic silicon—the increasing role of agriculture. Func Ecol. doi:10.1111/1365-2435.12544

Clymans W, Struyf E, Govers G, Vandevenne F, Conley DJ (2011) Anthropogenic impacts on amorphous silica pools in temperate soils. Biogeosci 8:2281–2293

Goldewijk KK, Beusen A, van Drecht G, de Vos M (2011) The HYDE 3.1 spatially explicit database of human-induced global land-use change over the past 12,000 years. Global Ecol Biogeogr 20:73–86

Guntzer F, Keller C, Meunier J-D (2012) Benefits of plant silicon for crops: a review. Agronomy for Sustainable Development 32, 201–213.

Liang Y, Nikolic M, Bélanger R, Gong H, Song A (2015) Silicon in Agriculture: From Theory to Practice. Springer, Dordrecht.

Strömberg CAE, Di Stilio VS, Song Z (2016) Functions of phytoliths in vascular plants: an evolutionary perspective. Functional Ecology, 30, 1286–1297.

Tansel B, Surita SC (2014) Oxidation of siloxanes during biogas combustion and nanotoxicity of Si-based particles released to the atmosphere. Environmental Toxicology and Pharmacology 37, 166–173.

Vandevenne F, Struyf E, Clymans W, Meire P (2012) Agricultural silica harvest: have humans created a new loop in the global silica cycle? Frontiers in Ecology and the Environment, 10, 243–248.

Chapter 8
Silica, Be Dammed!

It took us about 180,000 years but finally we did it. We hit Earth's carrying capacity for hunters and gatherers. That happened more or less 10,000 years ago and in order to keep going forth and multiplying, humanity had to learn how to farm. Talk about a multidisciplinary endeavor. There were plants and animals to be bred, tools to be designed, materials to be discovered, and a whole lot of biology, chemistry, hydrology, geology, meteorology, ecology, and biogeochemistry to be mastered. We're still working on it (and have added mechanization, transportation, refrigeration, genetic engineering, electronics, and information technology, among other things, to the list). Needless to say, our early stabs at farming were nowhere near as fruitful and reliable nor as intensive and destructive as the farming we do today. But as we slogged through the millennia, growing ever better at farming, ever more of us could be fed. So our numbers kept increasing. Do you see the vicious circle? As long as the human population keeps growing, so must the production of food through farming so that at least some chunk of the population that there has been enough food to produce doesn't then starve to death. For a long time, much of the getting better at farming meant increasing our control over the landscape and in no small part this was through damming. It also meant increasingly disrupting the biogeochemical cycles of nitrogen and phosphorus in our quest to keep cropland fertilized and productive. Both of these activities have had profound effects on the silica cycle.

8.1 To Put It in a Nutshell

Never mind the conversion of wilderness to farm and field that had begun some 5000 to 10,000 years previously, by about 5000 years ago (or 3000 B.C.E., shortly after the beginning of the Bronze Age), we really started to screw up the silica cycle. By this time, the total human population had reached perhaps 45 million people and the need to feed and water urban centers that had grown up out of ancient farming villages in parts of Asia and the Near and Middle East was severe. In response, we began to devise and deploy ingenious hydraulic technologies, such as canals, irrigation ditches, shadufs, chain pumps, wells, water wheels, and dams, for collecting, storing, diverting, and delivering water. Such measures changed the timing, energy, and paths by which water moves through and over land, altering not just the interactions between water and rocks and soils (and therefore the release of dissolved silica via silicate weathering), but the movement of water, solids, and nutrients (including silica) through rivers, ponds, lakes, marshes, and estuaries (and therefore the transport of dissolved silica to the sea).

Dams in particular have had huge effects on the biogeochemistry, ecology, and silica cycling of watersheds, creating lakes where there were not lakes before, trapping particles that would have otherwise been transported downstream, and obliterating seasonal flooding in favor of regulated year-round flow. While a few manmade dams would be nothing worth noting, we passed a few about 850,000 dams ago and more than 58,000 of these extant dams are technically large, exceeding 15 meters (49 feet) in height.[1] Altogether this means most rivers of any note have multiple dams upon them and clogging up their spider vein watersheds. This has had a massive effect on the silica cycle, taking a lot of silica entirely out of the game before it can be transported downstream to coastal waterways.

Meanwhile, eutrophication. Because we're good at adding insult to injury.

Eutrophication is the over fertilization of waterways with nitrogen (in the forms of nitrate, nitrite, ammonium, or dissolved organic nitrogen) and phosphorus (mainly as phosphate or dissolved organic phosphorus) but generally not silica via sewage and agricultural runoff. The eutrophication of inland and coastal waterways has increased in tandem with the number of people per hectare on Earth and with the amount of agricultural production of crops, meat, and dairy per hectare in a given area. Although in many places eutrophication is less severe than it was 20 years ago (because, guided by authorities such as America's Environmental Protection Agency, we've put work into cleaning up our act), if you are alive today and have never seen (or smelled) a eutrophied pond, river, lake, estuary, or coastal area in the throes of a pestilentially thick algal bloom, you've either been living in Antarctica or been in prison for most of your life.

Not only do these eutrophication-driven algal blooms clog up waterways, when bacteria, fungi, and herbivorous fish and invertebrates break down this material, they can use up most to all of the oxygen in a patch of water, making it a great place for fish and invertebrates to die. Dead zones, they're called, and they can be extensive. When the blooming algae are the sorts that produce toxins, many of which may be passed up the food chain and concentrated in fish and shellfish, there is the additional problem that the fish and shellfish become a health hazard to birds and humans and other mammals.

Worse yet, in our humble opinion as silica fans, nitrogen and phosphorus eutrophication frees up diatoms in lakes, ponds, and reservoirs to grow-grow-grow

---

[1] As defined by the International Commission on Large Dams (ICOLD).

8.1 To Put It in a Nutshell

and in so doing strip out incredible amounts of dissolved silica from the water. This is a major double whammy. This silica, now bound up in the beautiful frustules of biogenic silica that diatoms produce, ends up being buried in the sediments accumulating in lakes, ponds, and reservoirs instead of supporting diatom growth in estuaries and the ocean. That represents a serious break in the silica cycle that carried silica, weathered from silicate rocks, out to the ocean to support silica biomineralizers in the sea and the profundity of food webs based upon them.

8.2 A Brief History of Human Damming, or How Long Has This Been Going on

When did we begin to take control over the flow of water over the land? It's hard to say when we *first* built dams. On streamlet scales it must have started many thousands of years ago and undoubtedly even earlier in child's play. But our first building of *stupendous* dams is easier to demarcate. Massive structures leave signs that are hard for even raging water to fully wash away. So we think that the first of these impressive dams was Sadd-el-Kafar, a large embankment dam that the Egyptians began constructing around 4650 years ago (in 2650 BCE) at Wadi al-Garawi.

A stupendous dam this certainly was. Fourteen meters high, 100 meters (330 feet) wide at its base, and slightly more than 50 meters (165 feet) wide at its crest, Sadd-el-Kafara ran for 113 meters (370 feet) across the outflow of the wadi. The gargantuan, rubble-filled core at the heart of this dam was held in place by two interior walls, one on the upstream side and one on the downstream side of the core. These walls were made from precisely cut and fitted stones, identical rectangular cuboids each weighing 23 kilograms (about 50 pounds). Chunks of this structure remain standing today, giving testament to both how advanced we already were as architectural engineers and how much remained for us to learn.

A wadi, by the way, is a drainage course which is dry year round, except after rain, a point at which the wadi may play host to those ephemeral torrents of water known as flash floods. As you might imagine, the outflow of a wadi would have been a tricky place to build humankind's first massive dam.[2] While most of the time they were constructing the dam, they didn't have to deal with water because there wasn't any, they could not precisely predict when the next flash flood would hit, how much water there would be, nor how fast or violently the water would rise. And sure enough, 10 years into the construction of Sadd-el-Kafara, a particularly severe flash flood overtopped the dam, tearing it apart from the top down and carrying most of it away.

---

[2]Earth, on the other hand, has been making massive dams on her own out of ice and out of avalanches, etc., for as long as there has been landscape and water to run through it.

Live and learn, as they say. Now we build chutes, stepped channels, labyrinths, glory holes, side channels, ogee crests, and other spillways into our dams to pass excess water safely through or around the structure instead of letting it spill over the top in an uncontrolled manner. (Even with these extensive protections, overtopping remains a major cause of dam failure.) These developments happened much later, though, as people who sorely needed dams for one reason or another got tired of having them fail (at great trouble, expense, and loss of life).

Had it been completed, Sadd-el-Kafara could have held 500,000 cubic meters of water, enough to meet the needs of, for example, 400 typical modern American suburban households for one year. But Sadd-el-Kafara was not intended to create a reservoir for storing water. Yes, it was meant to be a plug, but only to protect downstream fields and settlements from floods like the one that did it in.

Over the thousands of years since Sadd-el-Kafara was constructed, the human race has been busier than beavers. At our dam-building peak post World War II, we were collectively constructing nearly *1000* large dams a year and perhaps a tenfold greater number of smaller ones (see Fig. 8.1 for the progression). As we don't take dams down at anywhere near the rate we construct them, we managed to accumulate the bulk of our current total of 850,000 dams in the mere 50 years of the second half of the twentieth century.

What are we doing with all these dams? Storing drinking water? It turns out, not just that, not by a long shot.

Most of the large dams we have built are *single-purpose dams*. Roughly half of these single-purpose dams have been built for the diversion of water for use irrigating fields. Not quite one-fifth have been built strictly to turn turbines to generate electricity. Only slightly more than a tenth were built to store water for residential, industrial, and agricultural use while only a tenth were built strictly for flood control. So profligate has our building of large dams been, more than 1 in 20 have been built *only* so that we may boat, swim, waterski, and partake in other what are known as lentic forms of recreation. That leaves also about 1 in 20 of large, single-purpose dams to have been built for other purposes entirely, including fish farming.[3]

Only 30% of the large dams we have constructed are multipurpose. Is this wasteful? Actually, it makes sense. You'd be working at cross purposes if you were trying to store water for agricultural, domestic, and industrial use while running it out through turbines to generate electricity.

---

[3]Small single-purpose dams can also be used to control the flow of water to mills, to store unwanted and generally hazardous byproducts of mining or other industrial activities, or even (in older times) to release water in a rush that rapidly transports a load of logs downstream and out of the forest.

8.2 A Brief History of Human Damming, or How Long Has This Been Going on

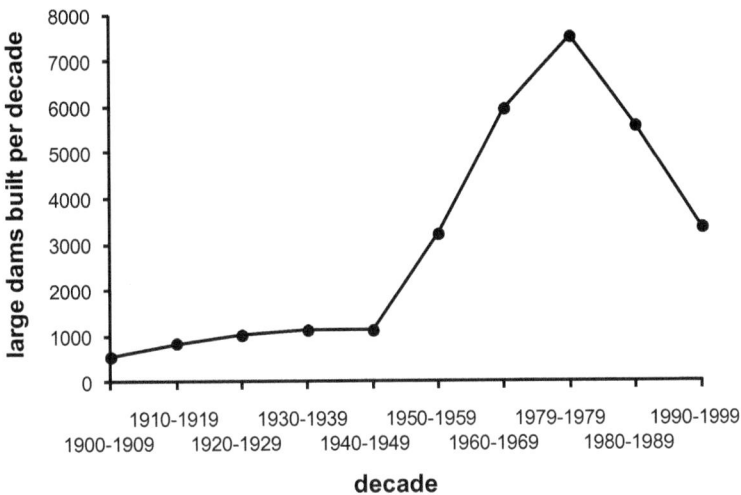

Fig. 8.1 Rate of large dam building per decade in the twentieth century. Data are from ICOLD world register of large dams

In any case, that late twentieth century dam-building frenzy means that most dams were constructed within the last hundred years and thus represent a recent, unprecedented, massive scale change that the surface of the Earth is still adjusting to.[4]

8.3 Dams and Silica

Even if the main purpose of many dams is not to store water, it is a common side effect. After all, dams curtail or even at times entirely block the flow of water through a generally narrow point in a river's run. The resulting pile up of water upstream of a dam widens the river, creates a new lake, or vastly increases the volume of lake that was already present. All of these possibility increase what is known as a river's total reservoir volume. They also increase the amount of time water spends in the area immediately upstream of the dam (as well as the amount of time water spends within the entire river system).

In physical terms, what happens when a cheerfully flowing river hits a reservoir is that the water slows down, the same as it does when a river runs into a lake. Particles, such as mud, sand, clays, and bits of biogenic silica such as opal

---

[4]It also means that a lot of these dams are reaching the end of their structurally safe age. Unfortunately, we haven't set aside money or made much in the way of plans to replace or remove many of them. Knowing the human race, that will probably have to wait until these dams start failing in droves.

phytoliths, siliceous sponge spicules[5], and diatom frustules, that were suspended in the energetic flow sink to the bottom where their fate is largely to be buried underneath the particles arriving after them. Because something like 65% of the world's freshwater flux encounters at least one dam on its way to the sea, there is, to put it mildly, an awful lot of previously riverborne material that is now sitting trapped behind a dam instead of continuing its travels toward the continental margin.

Reservoirs built in those cool, temperate zones that play host to much of Europe, Asia, and North America and therefore a large percent of the world's industrialized nations are the worst, retaining nearly half of this region's seaward sediment flux. Nearly half! This enormous retention of sediment occurs because there are a lot of dams in these regions and is made worse by cool, temperate zone rivers tending to be turbid (full of particles).

The worldwide average retention of sediment by dams is a lower but still jaw-dropping 20%. That is an awful lot of mud, silt, clay, and so forth that is not being transported to the coasts of the continents. Thus the most obvious effect of dams' the retention of riverborne particles is to deprive downstream areas of sediment, thereby causing or accelerating coastal erosion and the shrinkage of beaches.

Less obvious to the naked eye is the deprivation of downstream areas of dissolved silica. This deprivation occurs because a portion of the suspended material normally transported by a river dissolves en route, releasing dissolved silica into the river system to be delivered to the sea. But once particles are buried in a reservoir, sealed in their sedimentary tomb, there is little chance of this happening. This is one way that dams starve downstream areas of dissolved silica that would normally have been used to fuel the growth of diatoms, reeds and grasses, and other silica-producing organisms.

But there is a second process at work behind dams that is even more insidiously silica-stealing: diatom blooms.

When the moving water of the river hits a reservoir and slows down and all those particles that were in suspension sink out, the water becomes a lot more clear. This means light can penetrate into the water more than the couple of feet or inches it could before and that means photosynthetic plankton living in the water can suddenly make a good living. Phytoplankton can finally fix carbon into organic matter faster they respire it away. They can begin to grow.

But a dam means not only light, but also the time to put it to good use. Water that would have shot through that stretch of river in hours to days will now spend weeks to months to years in the extra reservoir volume. That's ample opportunity for phytoplankton like diatoms to build up biomass into thick blooms[6] and to remove almost all the dissolved silica in the water. And because these stretches of

---

[5]Yes, indeed, there are such things as freshwater sponges.

[6]They divide into two, then two becomes four, and four becomes eight, and so on like in the famous shampoo commercial from the 1970s until you're up into the millions of phytoplankton cells per liter, quite a lot of biomass.

8.3 Dams and Silica

quiet water with an enormously tall concrete wall at the downstream end are great places to build up sediments, the biogenic silica that has been produced stands a very good chance of sinking down and getting buried. The buck stops here, as they say, and as a result downstream areas are starved of silica.

Between sediment trapping and the reservoir effect on phytoplankton productivity, the 850,000 dams we have on their own would be more than enough to have a significant impact on the silica cycle. But even all 850,000 of them they probably wouldn't be enough on their own to drive downstream ecosystems into silica limitation, severe enough to stop diatoms from growing at maximal rates (or, indeed, from growing at all). That requires the help of eutrophication.

8.4 Dams, Eutrophication, and Silica

The reason that dams alone should not be enough to send downstream areas into silica limitation is that, naturally speaking, river waters tend to be rich in dissolved silica but poor in nitrogen and phosphorus. The silicate rocks that are weathering to provide solutes are mostly made of silica and have very little nitrogen or phosphorus in them. (Carbonate rocks also weather, but they are poor sources of all three of these nutrients.)

This natural poverty of nitrogen and phosphorus in freshwater normally prevents too much phytoplankton growth. For example, pristine rivers easily contain 100 micromoles of dissolved silica per liter, 160 micromoles of dissolved silica per liter, or even as much as 1000 micromoles of dissolved silica per liter, depending on how hot and wet the local climate is and how fresh and siliceous the geologic terrain. Meanwhile, there isn't usually more than a couple of micromoles per liter of nitrogen or phosphorus and this is just not enough. Freshwater diatoms tend to use six times as much silica as nitrogen, so that would mean that biologically available nitrogen concentrations would have to be an unrealistic 17 micromoles per liter for diatoms to use up 100 micromoles per liter of dissolved silica. Dissolved silica concentrations of 160 micromoles per liter (closer to the average river value) would require an even more unusually high nitrogen availability of 27 micromoles per liter. By the time you're in a highly siliceous tropical river containing 1000 micromoles per liter of dissolved silica, it would take 167 micromoles per liter of biologically available nitrogen for that dissolved silica to be entirely used up. And all these numbers assume the diatoms didn't have to share nitrogen with any non-diatoms. That would just make matters worse. And there would have to be a correspondingly sufficient amount of phosphate available, too.

Naturally speaking, it doesn't happen. There isn't that much nitrogen or phosphorus available. Diatoms growing in freshwater systems untouched by nitrogen and phosphorus pollution generally run out of nitrogen or phosphorus and stop growing long before they suck out all the dissolved silica, even if that water gets stuck for months and months behind a dam.

But natural is not the state of many rivers on Earth at this point. Never mind everything else we've done to them, for the last hundred or more years we've been continuously adding mind-boggling amounts nitrogen and phosphorus to rivers, groundwater, and lakes. The main culprits are fertilizers and animal waste flowing out of farms and off of fields, and poorly treated sewage (containing human waste and phosphate-containing detergents) from our houses, villages, towns, cities, and other settlements. As there is no equivalent addition of silica to balance things, the ratios of nitrogen to silica and of phosphate to silica in inland waterways have dramatically shifted against silica.

Thanks to our messiness, for the last few decades, diatoms in eutrophic systems *have not* been limited by nitrogen or phosphorus. They have been able to bloom until they have removed nearly all of the dissolved silica from the lake or river or pond or reservoir they are growing in. Some of this silica has recycled back into the water because some biogenic silica inevitably dissolves after the death of the diatom that made it. But a lot of the excessive biogenic silica that freshwater diatoms are now able to produce gets buried in reservoirs and lakes, preventing its delivery downstream to the sea.

Scientifically speaking, it took us some time to notice that dissolved silica was disappearing and yet some more time to grasp why. Of course, in retrospect, it's totally obvious. Of course this is what happened when we overloaded waterways with nitrogen and phosphorus. But in the beginning, we were probably too shocked by the eutrophication-fueled overgrowth of phytoplankton in general and all of the clogging and fouling of waterways and all of the fish-killing it was doing. Plus who would expect excessive nutrient addition to result in nutrient loss?

And hardly anyone had the cleverness to foresee that dams would sequester silica.

It took study of three different systems over an embarrassingly large number of decades for us to figure out what has been going on.

8.5 Case Study #1: The Laurentian Great Lakes

The first case that came to light of how we're screwing up the silica cycle has nothing to do with a dam, but strictly with eutrophication. It was also our first inkling that freshwater ecosystems were being shifted into silica limitation as a side effect of all the phosphorus and/or nitrogen we were spilling into waterways.

The time was the 1970s. Two to-be-giants in the field of limnology[7], Claire Schelske and Eugene Stoermer[8], both of the University of Michigan, had been

---

[7]Limnology is the study of inland waters, including rivers, ponds, lakes, reservoirs, wetlands, estuaries, and groundwater, with focus on the interactions between organisms and their environment.

[8]Incidentally, Eugene Stoermer was also the co-namer of the Anthropocene.

8.5 Case Study #1: The Laurentian Great Lakes

studying the Laurentian Great Lakes that lie along the US-Canadian border. Before they started this work, a series of measurements on the intake waters of filtration plants serving the city of Chicago had shown that dissolved silica concentrations in Lake Michigan were decreasing. Claire and Eugene quickly discovered that the situation had escalated to the point where diatoms in Lake Michigan were running out of dissolved silica before the end of summer. This caused the growth of diatoms to screech to a halt several months before the end of their natural growing season, which had previously extended into autumn. Because diatoms serve as the base of key food webs, the knock on effect of premature stoppage in their growth was food shortage for fish and invertebrates in autumn in Lake Michigan.

One of the first things Claire and Eugene wondered was whether the mid-summer silica depletion was something new or if it had merely previously escaped notice. The two of them sleuthed through what old, patchy datasets they could dig up from various water quality agencies. The resulting data, plotted in Fig. 8.2, revealed that the mid-summer exhaustion of dissolved silica was new. It had first showed up in Lake Michigan and settled itself fully in sometime between 1955 and 1969.

Now all they had to do was figure out why it was happening.

At first it must have been a head-scratcher. The decades leading up to and including the one that they were in had seen an explosive growth of algae beginning

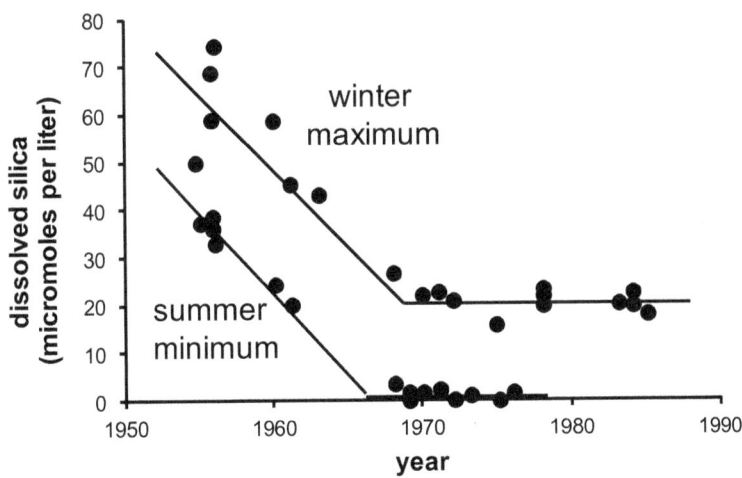

Fig. 8.2 The typical maximal concentrations of dissolved silica in winter and the typical minimum concentrations in summer in the surface waters of Lake Michigan significantly decreased during the 1950s and 1960s. This was due to excess production of biogenic silica fueled by phosphorus eutrophication. This figure has been redrawn from Internationale Revue der gesamten Hydrobiolie und Hydrologie 73, Schelske CL, Historic trends in Lake Michigan silica concentrations, 559–591, (1988), copyright © 1988 WILEY-VCH Verlag GmbH & Co. KGaA, Weinheim, with permission from Wiley

to choke freshwaters across the globe. Diatom populations hitting the wall due to silica scarcity flew against the grain of this trend. It was out of step with everything turning most rather unexpectedly, unprecedentedly green.

But, when Claire and Eugene sat down and thought about it, the answer became obvious. There was a growing understanding, all over the world, that lakes, ponds, rivers, and estuaries were clogging up with excess algal growth because their concentrations of nitrogenous and phosphatic nutrients (that all phytoplankton, including diatoms, need in order to grow) were going off the charts. One reason, very well accepted at that point, was agricultural runoff that was full of biologically utilizable nitrogen and phosphorus from all the manure produced by livestock and from fertilizers applied to fields. The other widely acknowledged cause for eutrophication was human-generated sewage. But neither of these things, though, was adding much in the way of dissolved silica.

What Claire and Eugene also quickly came to realize was that, in the Great Lakes' case, the problem wasn't so much one of agricultural runoff nor was it one of too many people generating too much sewage. The problem was detergents that contained a lot of phosphate.

Detergents are crafty chemicals. One side of them is good at binding oils and fats (which, on their own, are insoluble in water) and another side of them is good at being dissolved in water. When detergents bind an oil or a fat, they thus drag it into solution. That is their cleaning power.

But detergents are not as effective in hard water, which is water that has a lot of calcium and magnesium dissolved in it. The doubly charged cations of Ca^{2+} and Mg^{2+} bind to detergents, precipitating them. Instead of foaming and doing a spanking great job of cleaning, in hard water, detergents form stubborn scum.

Most store-bought detergents have a water softener included in them to chemically preoccupy Ca^{2+} and Mg^{2+} so the detergent can do what a detergent's gotta do. Starting from about the 1950s and still in many cheaper detergents sold today, that water softener has been a phosphate such as trisodium phosphate or sodium hexametaphosphate (pronounced hexa-meta-phosphate).[9] Washing such phosphate-containing detergent down the drain, especially in areas lacking effective sewage treatment facilities, drains that phosphate straight into the nearest lake, river, or ocean, where it can feed the algae.

Take yourself back to the middle years of the twentieth century, when the human population was just beginning to embark upon the steep, wild, and crazy part of its exponential increase. At the same time modernization was marching along, introducing the washing machine and the dishwasher and then boosting their use. So not

---

[9]These days we are tending towards using zeolites instead because they don't add massive amounts of a major nutrient to the water, although by even this very late date, there are few *national* laws against the use of phosphate in detergents.

8.5 Case Study #1: The Laurentian Great Lakes

only were there vastly more people than ever before living along the shores of the Great Lakes, flushing toilets and generating agricultural runoff, they were also enthusiastically doing laundry and washing dishes and the detergents they were using consisted of up to 50% phosphate.

There is only so much even a Great Lake can take, even if it is the fifth largest lake in the world. For Lake Michigan, this was the waste water from millions of loads of laundry and dishes done each week by the people of Chicago, Milwaukee, Green Bay, and other sites on the shore on top of all the other sewage and agricultural runoff. Algae began growing like gangbusters.

As we've said before, Claire Schelske and Eugene Stoermer realized that this phosphorus eutrophication was the key to the disappearance of the silica from Lake Michigan. With the phosphorus brakes released on their growth, diatoms in Lake Michigan could grow until they had converted basically all of the dissolved silica in the lake's sunlit surface waters into diatom frustules that ended up in the sediments.

In other words, dissolved silica was disappearing from Lake Michigan because it was being turned into particulate silica by diatoms growing in the surface waters of the lake and exported, via sinking, to the bottom of the lake.

Ecologically speaking, this was dire news, but intellectually it was kind of cool. Phosphorus eutrophication was leading to silica oligotrophication, a paradox that made perfect sense. And, as ideas go, it was one that Claire and Eugene could test.

Even long-lived lakes are only temporary features of the landscape; they're busy filling up with layer upon layer of sediment. You can take advantage of this if you want to learn about the history of a lake, ecologically and climatically speaking. All you need to do is carefully take a sediment core, slice it lengthwise in half, and, moving downwards from the top of the core, begin your journey back through time.

The core that Claire Schelske, Eugene Stoermer, and their collaborators took in the middle of the deepest part of Lake Michigan (which is found within Grand Transverse Bay) was 40 centimeters long (about 16 inches) and, based on radiometric dating[10], covered the last century and a half. The milligrams of biogenic silica to be found per gram of core versus depth within the sediment core are shown in Fig. 8.3 and that's all that is needed to show the story.

Before 1920 or 1930, only 10 milligrams of every gram of sediment that was accumulating on the lakebed was biogenic silica, a content that can also be expressed as 10 parts per thousand, or 1% biogenic silica. As biogenic silica accumulation rates go, that's pathetic. Those pre-postmodern Lake Michigan diatoms should hang their heads in shame at the poor job they were doing of exporting silica to the sediments. Or maybe they should be proud, because this is the level of export of silica that the lake could maintain, given the amount that was being delivered to it each year in runoff.

---

[10]Using the isotope lead-210 (^{210}Pb), which has a half-life of 22 years and is a particulate material which is continually falling out of the atmosphere following its production by the decay of radioactive radon gas.

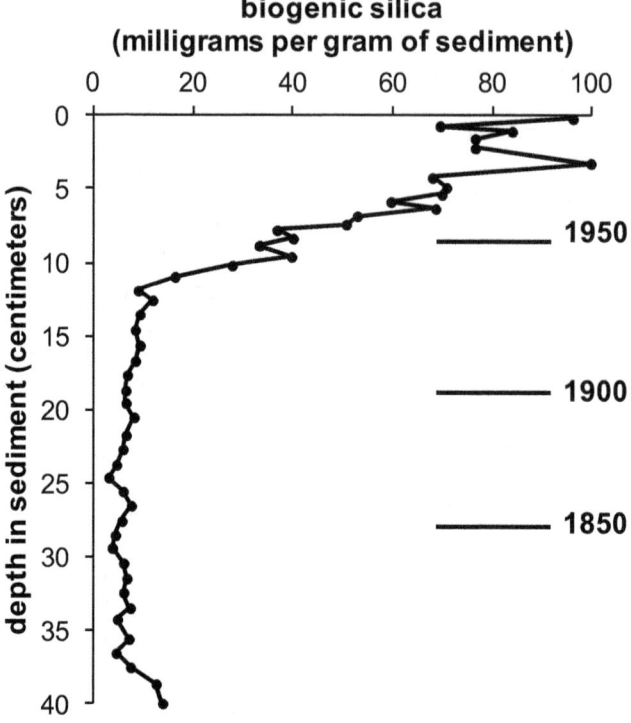

Fig. 8.3 Biogenic silica concentration versus depth in a core from Grand Traverse Bay, Lake Michigan records the transfer of Lake Michigan's silica into the sediments during the twentieth century's explosion in the use of phosphate detergents and fertilizers. This figure has been redrawn from Hydrobiologica 143, Schelske CL, Conley DJ, Stoermer EF, Newberry TL, Campbell CD, Biogenic silica and phosphorus accumulation in sediments as indices of eutrophication in the Laurentian Great Lakes, 79–86, (1986), copyright © Dr W. Junk Publishers, Dordrecht, with permission from Springer

As you move forward in time in the sediment core from 1930 it is like running up a ramp. The biogenic silica content of the sediments steadily climbs, reaching a peak of 100 milligrams of biogenic silica per gram of sediment (or 10% biogenic silica) by 1970. This you can see clearly in Fig. 8.3. What Fig. 8.3 does not show is what you would see if you made this measurement on lake sediments stretching back to the birth of Lake Michigan. The biogenic silica content of the older sediments (not shown here) reveal that such an astonishing change in silica burial had never happened before, not once in Lake Michigan's 14,000 year history.

8.5 Case Study #1: The Laurentian Great Lakes

Thus Claire and Eugene had managed to show that the disappearance of dissolved silica was due to excessive production of biogenic silica and that this excessive production of biogenic silica had never happened before people moved in by the tens of millions and started doing a lot of laundry in cities and towns along the shore of Lake Michigan.

There is another big detail of the shift that is revealed by the sedimentary record —the shift in the composition of the diatom population with the eutrophication of Lake Michigan. Diatom frustules are, after all, quite distinctive, and can be used to tell one diatom species from another.

Sediments older than the expansion of European settlements around Lake Michigan in the 1800s had a lot of different diatom species in them, representing a wide range of ecological niches (high light, low light, high silica, low silica, benthic, planktonic, spring growing, summer growing, autumn growing, etc.). But as phosphorus eutrophication (and silica depletion) increased, the variety of diatoms in the sediments narrowed to only those species that grow in late winter/early spring. This is the very start of each year's growing season, when dissolved silica concentrations are still high from winter mixing. Diatoms that would have grown later in the year were missing because by the time it was their turn to grow, there was no dissolved silica left for them to use. Benthic diatom species, meaning those that grow on the bottom of the lake (in shallow waters), also disappeared, most likely because the overgrowth of plankton due to eutrophication made it too dark down there for them to grow.

Overall, diatoms getting shut out of the latter part of the growing season in Lake Michigan while there is still plenty of nitrogen and phosphorus available for growth is a bad thing. It means a decrease in the flow of energy and materials through diatom-based food webs, which generally efficiently lead to fish, and an increase in the growth of noxious plankton species like dinoflagellates.[11] Worse yet, what happens in Lake Michigan doesn't stay in Lake Michigan. Now stripped of their dissolved silica, the waters of Lake Michigan flow into Lake Huron and then Lake Erie, go over Niagara Falls, flow into Lake Ontario, and then via the Saint Lawrence River, arrive at the Atlantic Ocean at the Gulf of Saint Lawrence in all the full glory of their silica deficiency. You can almost hear the coastal diatoms screaming.

---

[11]The addendum here is that the water quality (and dissolved silica content) began improving in the first decades of the twenty-first century due to improvements in sewage treatment and to the phasing out of phosphate detergents. Then the quagga mussel invaded, via larval stages that most likely arrived in water released from the ballast tanks of transoceanic shipping vessels. The quagga and its relatives are voracious filter feeders and they've colonized enough of Lake Michigan to keep the waters clear of algal blooms, regardless of the lake's nutrient status. Unfortunately, this means that phytoplankton still aren't making it into the food chains that lead to fish, causing a collapse in the lake's fisheries. Poor Lake Michigan can't catch a break from the trouble caused by human beings.

8.6 Case Study #2: The Baltic Sea

Once the research on Lake Michigan became known, people started to have a look at other large inland bodies of water to see if the same things were happening to their local silica cycles. One of the more recent and most intensively investigated places has been the Baltic Sea.

Lake Michigan is big, but the Baltic Sea is massive. It is that large inland sea that sits in the midst of northern Europe. You could think that the Baltic Sea is too big to be notably affected by the activities of humankind. But once you start paying attention, you quickly come to pity the Baltic Sea. All but entirely encircled by Sweden, Finland, Russia, Estonia, Latvia, Lithuania, Poland, Germany, and Denmark and additionally containing portions of Belarus, Ukraine, Norway, Slovakia, and the Czech Republic in its watershed, the Baltic Sea is subject to continual insult by the agriculture and sewage of 90 million people. This insult, which comes partly in the form of four to eight times more nitrogen and phosphorus than it tended to receive a century ago, is delivered in 16,000 metric tons of freshwater *per second* flowing off the land. Consequently, concentrations of nitrate and phosphate in the Baltic Sea have increased over the last century.

But concentrations of dissolved silica have declined. This decline has been severe. For example, concentrations of dissolved silica in subsurface waters in a central area of the Baltic Sea have decreased by one-third to two-thirds since the late 1960s, the time when monitoring began at that location.

So far so Great Lakes? It seems pretty similar. Just add eutrophication and watch those diatoms go (until they run out of dissolved silica).

Maybe. But maybe not.

There is a certain key difference in the situation of Lake Michigan and the Baltic Sea. Excess phosphorus is mainly delivered to Lake Michigan directly, from sources that originate along the shore. But the Baltic Sea is receiving waters high in phosphorus and nitrogen via rivers that travel great distances to get to the Baltic Sea and generally encounter multiple lakes and dams along the way. This gives silica plenty of opportunity to be removed and trapped in sediments long before it arrives into the Baltic Sea itself. So maybe excess production of biogenic silica within the Baltic Sea itself is stripping dissolved silica out of its waters. But maybe that silica is being removed upstream and because the Baltic Sea is being thus deprived of dissolved silica, its poor diatoms aren't growing (or producing biogenic silica) much at all. If we want to help solve the problem (in part because we'd like to get the Baltic back to supporting food webs that produce something besides enormous swarms of jellyfish) we need to know which one is going on. Plus we're just plain old curious.

The first question to tackle: do outputs of silica from exceed inputs of silica to the Baltic Sea? If so, at least some of the problem is due to eutrophication-fueled diatom growth within the Baltic itself. The straightforward way to answer the question is to put together a silica budget with inputs on one side and outputs on the other.

8.6 Case Study #2: The Baltic Sea

For a small reservoir or lake, this should be easy. You need to measure three things. One is the amount of dissolved silica flowing in with water flowing in from rivers and streams. Another is the amount of biogenic silica accumulating on the lake bed. The third is the amount of dissolved silica flowing out in the stream that serves as the lake's outflow.

But the Baltic Sea is no little lake. Nearly 100 rivers of note flow into the Baltic Sea and you'd have to monitor each one for several years. There is at least only one outflow of water from the Baltic Sea (aside from evaporation): water leaves via the Denmark Straits to the Atlantic Ocean. But sometimes, because of storms, winds, currents, and tides, the water flows in instead, bringing dissolved (and biogenic) silica with it. As far as measuring how much biogenic silica is getting buried in Baltic Sea sediments, the complication here is that the Baltic Sea is made up of numerous basins, such as Bothnian Bay, the Bothnian Sea, the Gulf of Finland, the Baltic Proper, the Gulf of Riga, the Denmark Straits, and the Kattegat, and they all behave differently (and exchange water with each other). Figuring out how much biogenic silica is accumulating in the sediments requires careful study of sediment accumulation rates (and correction for sediment winnowing and focusing due to currents) in all of these regions.

Despite the near impossibility of the task of determining whether more silica is leaving the Baltic Sea than is coming in, there have been several attempts to put together silica budgets for the Baltic Sea. (Scientists do love them a challenge.)

In their quest, two different groups of researchers have fed monthly measurements of dissolved silica from the major rivers flowing into the Baltic and measurements of wintertime dissolved silica concentrations at various locations within the Baltic Sea into a computer model of Baltic Sea circulation in order to calculate how much dissolved silica is disappearing from Baltic Sea water as it flows out to the North Atlantic. Both groups came up with much the same result, that recently roughly 1,300,000 tons of silica has been accumulating in the sediments of the Baltic Sea each year. As both modeling efforts produced a Baltic Sea whose dissolved silica concentrations decreased from year to year during the model runs, an export of 1,300,000 tons of biogenic silica to the sediments must be enough for the total export of silica from the Baltic Sea to be exceeding silica's input.

So the Baltic Sea is probably at least a little bit like Lake Michigan. Eutrophication is causing it to overproduce biogenic silica. But when you actually look at the data that were fed into the models, you realize that something else pretty major is going on.

Many of the major rivers flowing into the Baltic Sea have concentrations of dissolved silica that are, frankly, shocking.

The Neva River, which is the greatest of the rivers flowing into the Baltic Sea, contains *8* micromoles of dissolved silica per liter when it reaches the Baltic Sea. Can you hear the diatoms weeping? No self-respecting river should contain such a measly amount of dissolved silica. An *average* (as in mediocre, hum-drum, run-of-the-mill) river has 160 micromoles of dissolved silica per liter and an overachiever has 1000 micromoles of dissolved silica per liter. A number like *8* is almost unfathomable. To yield up only 8 micromoles per liter dissolved silica, the Neva River's catchment is only producing a net 63 kilograms of silica per square kilometer of catchment area per year, another number to make a diatom cry.

The Vistula, the Baltic Sea's number two river in terms of the delivery of water, is better, but at 119 micromoles of dissolved silica per liter and 580 kilograms of silica produced per square kilometer, still below average. The number three river, the Daugava, averages around 60 micromoles of dissolved silica per liter, a yield of 411 kilograms of silica per square kilometer, more dismal numbers.

But if you look at the rivers draining into the Baltic Sea from the emptier, more northern areas of the catchment, you'll find that they are not like this. The Närpiönjoki in Finland has an average dissolved silica content of 267 micromoles of dissolved silica per liter, representing a catchment yield of 2285 kilograms of silica per square kilometer. The numbers for the Isojoki, also in Finland, are similar: 195 micromoles of dissolved silica per liter and 2105 kilograms of silica per square kilometer.

What's the difference between the respectably silica-containing rivers and the failures? The silica-poor rivers draining into the Baltic Sea are found in more heavily populated areas while the silica-rich rivers are running wild. The silica-poor rivers are suffering from notably greater eutrophication and they contain much greater (natural and manmade) reservoir volume.

Take that astonishingly low-silica river, the Neva River, for example. Just upstream of St Petersburg (not so far from the Baltic shore), it runs through Lake Ladoga, one of the largest lakes in Europe. Lake Ladoga has been heavily eutrophicated since the 1960s. You can all but walk on the phytoplankton blooms, they grow so thick. This is where a lot of the Baltic Sea's silica is ending up. Buried in Lake Ladoga's sediments.

Similar, although less severe losses of silica must be occurring in lakes and reservoirs along other eutrophicated rivers that feed into the Baltic Sea.

Once you have data (on silica concentrations, water flows, surface area of river catchments, and so on), you can cross-examine them to tease out the combined effects of eutrophication and damming on the silica content of rivers draining into the Baltic Sea. You could, for example, plot the concentration of dissolved silica in a river (or, if you'd prefer, its yield of dissolved silica per catchment area) versus the amount of time water spends in the river's catchment area. Dams and natural lakes both increase the residence time of the water within the catchment. Thus a long residence time indicates the water spends a lot of time in places favorable to diatom blooms and export of silica to sediments.

In practice, residence time of water in a river catchment is not an easy thing to measure. So you can try to use a proxy, some other more easily or accurately measurable factor that is relatable enough to residence time that it can serve as a stand in. You might try hydraulic load, the amount of water, expressed as meters of height, that passes over a point in the river system each year. High hydraulic loads

8.6 Case Study #2: The Baltic Sea

are associated with short residence times (fast flowing water) while low hydraulic loads indicate long ones (fairly slow, stagnant flow).

The result? The lower the hydraulic load (and the longer the residence time of water in the river system), the lower the catchment's dissolved silica yield per area and the lower the concentration of dissolved silica in the river. This is true for all types of river feeding into the Baltic Sea, meaning that reservoir volume (be it natural lakes or manmade due to damming) is giving diatoms a chance to bloom and remove silica before the silica reaches the Baltic Sea. That the problem is worse in eutrophicated river systems is also clear because concentrations of dissolved silica are lower in these rivers regardless of their hydraulic load.

This is all illustrated nicely in Fig. 8.4. The yields of dissolved silica per catchment area from a subset of the Baltic Sea catchments that are not eutrophicated and whose flows are not interrupted by dams (represented on the plot by the black triangles) range from 800 to almost 1300 kilograms of silica per square kilometer. These highest silica yields belong to fairly pristine catchments where water doesn't spend too much time hanging around. There is neither the time nor the added nitrogen and phosphorus for diatoms to bloom and remove dissolved silica. These rivers hit the Baltic Sea with a healthy load of dissolved silica and this most likely represents the natural state of the system.

The subset of the Baltic Sea catchments that are not eutrophicated but are subjected to damming (represented by the black circles on Fig. 8.4) give yields of 280 to 1100 kilograms of silica per square kilometer and the yields clearly decrease as the residence time of water in the watershed increases because waters are detained in lakes and reservoirs on their way to the Baltic Sea. There is just enough naturally occurring nitrate and phosphate for diatoms to bloom and remove some dissolved silica from the river system, thus preventing it from reaching the Baltic Sea.

The subset of the Baltic Sea catchments investigated that are both eutrophicated and dammed (the gray circles on Fig. 8.4) have yields of 60 to 600 kilograms of silica per square kilometer. There are two things going on here. Eutrophication in general is keeping the yields low because diatoms are growing and removing silica. But the lowest of the low yields are occurring in catchments where the residence time of the water is the greatest and this is because of lakes and by reservoirs produced by damming.

These are, you might say, damning results for both eutrophication and damming.

Worse, we're starting to suspect that there is another process contributing to the problem. Rates of silicate weathering within damned catchment areas are probably lower than in non-dammed catchments. When you build a dam, a lot of what used to be the soils of grasslands and forests becomes the bottom of a reservoir. Weathering reactions tend to be vigorous in the soils of grasslands and forests and the dissolved silica the weathering produces is efficiently flushed into rivers when it rains. But not much silicate weathering is going to be going on at the bottom of a reservoir. As a result, not only is the reservoir sequestering silica that used to flow onwards to the ocean, it is also preventing some silica from being added to the river in the first place.

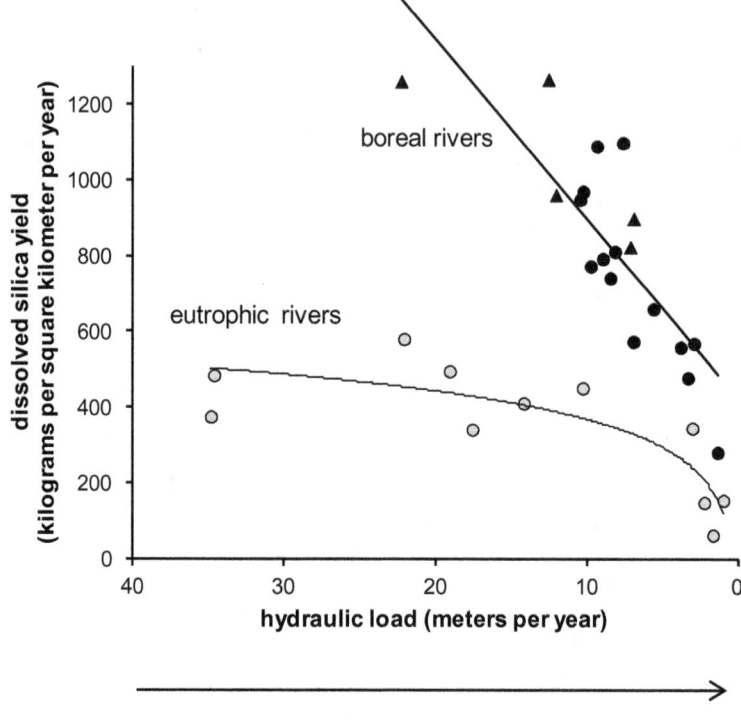

Fig. 8.4 Yield of dissolved silica compared to hydraulic load for pristine boreal rivers (*black triangles*), eutrophicated but undammed boreal rivers (*black circles*), and heavily eutrophicated and dammed rivers (*gray circles*) draining into the Baltic Sea. Hydraulic load serves as a proxy for residence time, with residence time increasing to the *right* of the figure. This figure has been redrawn from Journal of Marine Systems 73, Humborg C, Smedberg E, Medina MA, Mörth C-M, Changes in dissolved silicate loads to the Baltic Sea—the effects of lakes and reservoirs, 223–235, (2008), copyright © 2007 Elsevier B.V., with permission from Elsevier

The patterns revealed by the set of river catchments of Fig. 8.4 can be used to estimate that the combination of eutrophication and damming has decreased the amount of dissolved silica delivered into the Baltic Sea by 30–40% during the last century.

That is the sort of number that you should not announce to a silica enthusiast unless you've taken the precaution of sitting them down in a chair.

In other words, right now is a really lousy time to be a Baltic Sea diatom. All those dams we've built within the Baltic Sea catchment area and all the nitrogen and phosphorus we're pouring into it too… We're denying Baltic Sea diatoms the silica they need to survive.

8.7 Case Study #3: The Black Sea

The same sort of thing has been going on in the Black Sea. But it took us some time to understand the problem here because we got hung up on the idea that the Black Sea's silica problems stemmed from one single spot on the Danube River. It was at the time a compelling story, though, and it got a lot of people interested in the problem of silica and dams.

The Danube River, the second longest river in Europe, arises in the Black Forest, meanders through or along Germany, Austria, Slovakia, Hungary, Croatia, Serbia, Bulgaria, Moldova, Ukraine, and Romania, collecting water from an even greater portion of Europe, and then pours into the Black Sea. Like the also enormous catchments of the two other major rivers that flow into the Black Sea (the Don and the Dieper), the Danube River catchment was dammed to the hilt between the 1960s and the present day and eutrophication is a serious problem.

At this point, you will probably nod knowingly when we tell you that for decades phytoplankton overgrowth in the Black Sea has been common and concentrations of dissolved silica have declined and that all of this has occurred alongside shifts in the composition of photosynthetic populations away from diatoms. Populations of benthic macrophytes such as seagrasses and macroalgae have also taken a serious hit and anoxia has become more widespread. This is bad, not only for Turkey's Black Sea anchovy fishery, but also for entire Black Sea beach resort industry that is important to the economies and vacations of Turkey, Bulgaria, Romania, Ukraine, Russia, and Georgia.

By 1997, which was long after Claire Schelske's and Eugene Stoermer's groundbreaking discovery that eutrophication resulted in silica removal within the Laurentian Great Lakes but a few years before the work on the Baltic Sea described in the previous section, it became clear that the Black Sea was missing a lot of dissolved silica. Monitoring that had begun in 1960 had revealed that dissolved silica concentrations had briefly shot up and then steadily dived in the Black Sea, dropping from an average of about 60 micromoles of silica per liter in the mid-1970s down to below 10 by about 1995. It was as if some single abrupt change in the system kicked off a dramatic disappearance of silica.

Enter the Iron Gate dam.

Iron Gate I runs across the Danube along a stretch that serves as the Romanian–Serbian border and is a hydroelectric generation plant. It's also the largest dam on the Danube River. It was so very most suspiciously completed in 1972, right before dissolved silica went into its dive. The obvious explanation was that the dissolved silica that was now missing from the Black Sea was accumulating as 1.3 million kilograms of biogenic silica accumulating in the sediments trapped by Iron Gate I every year. It had to be the case! The circumstantial evidence was screaming.

Simple stories are seductive. Impound a single dam on a major tributary of a fair-sized sea and send the downstream coastal areas into silica freefall. Even scientists can get enthusiastic over ideas that are too good to be true.

Of course the story turned out to be wrong, but it also turned out to be the story that got researchers interested in the effect of dams on silica. Although the problem of dams resulting in downstream waters depleted of dissolved silica had first come to light in a well-researched and well-written 1980 publication in a top-notch scientific journal by top-notch scientific researchers[12], somehow the news hadn't sunk in.

The dam's name probably helped too. With a name like that (Iron Gate) the dam had to be: Huge! Solid! Fearsome! Authoritarian! And rather like the Iron Curtain. Nothing, not even silica, could get through. Even silica scientists (initially) felt swayed. Damn that dam for stealing all the Black Sea's silica.

However, the name was merely an accident of history. Iron Gate I and its younger sibling, Iron Gate II, were not named for their formidability. They were named for their locality.

The Iron Gates are a picturesque series of gorges that the Danube runs through. That narrowed stretch of river is over a hundred miles long and its name may have originated from a number of pestilential bedrocks shoals, now long since removed, that could rip apart the hulls of ships that failed to steer around them. Perhaps hundreds of years ago, a person with poetry in their heart likened them to the spikes on an iron gate. In an alternate universe, one with a shortage of poets, there may be two Damned Shoals dams instead and nobody who thinks that particulate silica is piling up behind them.

In this universe, at least, biogeochemists eventually decided to put the hypothesis to the test.

One of the first things they noted was that, yes, the reservoir upstream of Iron Gate I is nothing to sneeze at. It is 120 kilometers (75 miles) long and holds 2.4 billion cubic meters of water. But this has more to do with the river than the dam. The Danube River carries so much water by the time it reaches the Iron Gates, the Iron Gate I dam needs 1100 meters of length in order to span the river. That's 3600 feet. So it doesn't take much diminishment in flow (or increase in water residence time) to create a large reservoir volume at this location.

Indeed, the Iron Gate dams do not significantly increase the residence time of water along this stretch of the Danube. Iron Gate I, like Iron Gate II, is a hydroelectric power generating station. Water flows FAST through the dam. It has to turn turbines.

When you measure how long the water spends in this reservoir (plus in the much smaller one downstream of it now that they've built Iron Gate II), you find that it's six and a half days, which is hardly any time at all. If we were a diatom bloom, we'd file a complaint. It's not enough time to get our work done.

---

[12]Larry Mayer and Steven Gloss, two widely known and respected biogeochemists, had first noted the effect on dams on dissolved silica in 1980 in a published paper on the Colorado River in Arizona before and after the construction of Edward Abbey's favorite of favorites, the Glen Canyon Dam.

8.7 Case Study #3: The Black Sea

Silica budgets constructed for Iron Gate I, its reservoir, and its sediments have confirmed this. The Iron Gate dams are not a big trap for silica. It looks like about 850,000 tons of silica flows into the reservoir as dissolved silica each year and 810,000 tons flows out. The 40,000 tons of silica trapped is not nothing but it is well short of the postulated 600,000 tons.

Now that the research on silica losses in the Baltic Sea catchment area has been done, it's clear that the disappearance of silica from the Black Sea isn't the work of one fearsome dam. It's due to the tens of thousands of dams that have been built along the often eutrophicated rivers that head ultimately to the Black Sea. And of course, the eutrophication-fueled overgrowth of diatoms within the Black Sea itself has probably contributed to the decades-long decline in dissolved silica concentrations that has occurred within the Black Sea. This isn't as superficially exciting a story as one single dam having an impact so severe it was changing the ecology and biogeochemistry of the entire Black Sea, but it is profound.

8.8   The Global View

The Larentian Great Lakes, the Baltic Sea, and the Black Sea are all major bodies of water, but they are still just a drop in a bucket compared to all the fresh and salt water on Earth. But they represent a crisis that is unfolding across the Earth.

Those 850,000 dams we currently have, about 60,000 of them large, disrupt flow *on more than half* of all river systems. Various different modeling studies have calculated that altogether these dams decrease the total global flux of dissolved silica to the ocean by 5%.

Does that seem small?

It isn't. Five percent represents a deficit that adds up year after year into enough of an enormous lack of dissolved silica that diatoms in estuaries and the coastal ocean, where they are most needed to support fisheries, could be operating at a disadvantage.

Natural but now eutrophicated lakes are even worse than reservoirs, in that they seem to be better at retaining silica in their sediments. Including them in the global estimate would bring the total amount of silica flux diminished to nearly *30%*, a truly apocalyptic number.

The next time you're down at the beach, you may want to hug a diatom. They could certainly use the moral support. But they'd like it even more if you pushed your legislatures for laws requiring that detergents to be sold phosphate-free. This needs to include dishwasher detergents (which, in many places, are exempt from restrictions). Our diatom friends could also use more and better sewage treatment plants, incentives and support for farmers to do something about the millions of tons of phosphate and nitrogen fertilizers and animal waste flowing off of their lands, the removal of dams no longer needed, and some serious, sober second thoughts about building any more.

Further Reading

Conley DJ, Stålnacke P, Pitkänen H, Wilander A (2000) The transport and retention of dissolved silicate by rivers in Sweden and Finland. Limnol Oceanogr 45:1850–1853

Humborg C, Conley DJ, Rahm L, Wulff F, Cociasu A, Ittekkot V (2000) Silicon retention in river basins: far-reaching effects on biogeochemistry and aquatic food webs in coastal marine environments. Ambio 29:45–50

Maavara T, Dürr H, Van Cappellen P (2014) Worldwide retention of nutrient silicon by river damming: from sparse data set to global estimate. Global Biogeochem Cycles 28:842–855

Schelske CL, Stoermer EF, Conley DJ, Robbins JA, Glover RM (1983) Early eutrophication in the lower Great Lakes: new evidence from biogenic silica in sediments. Science 222:320–322

Syvitski JPM, Vörösmarty CJ, Kettner AJ, Green P (2005) Impact of humans on the flux of terrestrial sediment to the global coastal ocean. Science 308:376–380

Chapter 9
The Venerable Silica Cycle

For as long as it has had both water and rock, the Earth has had a silica cycle. This cycle begins with the creation of the silicate rock of new mountains, mountain belts, and ocean floor. Weathering wears these materials down, transferring their silica slowly to the sea. Thanks to plate tectonics, solid silica that ends up at the bottom of the ocean returns to the interior of the Earth to replenish the reservoir of material from which new silicate rock is made. These days, along the way from rock to sea to deep sea sediment, silica cycles through the fields and forests, diatom blooms, and glass sponges of the biosphere. But these enormous diversions are a new thing. For most of Earth history, there was no such thing as silica biomineralization and there were no biological loops in the silica cycle. Thus the silica cycle that we have today, the one that we tend to think of as normal, default, and everlasting, has not just features but *major features* it did not possess for the majority of its existence. On the other hand, that shouldn't surprise us. That's what happens with systems that progressively evolve.

9.1 The Silica Cycle

To a biogeochemist, the surface of the Earth (including the atmosphere and upper mantle) is a place where things cycle from one reservoir to another all the way around until they arrive at the beginning and start over again. Even more interestingly, the major biogeochemical cycles, those of carbon, nitrogen, phosphorus, sulfur, oxygen, and silica, interact with each other, with climate, with the hydrologic cycle, and, of course, the biosphere. There is a good and intellectually satisfying living to be had studying one or more of these cycles. How do they work? Who participates in them? How severely they have been perturbed by the recent rapid rise in the number of human beings and all the crazy things that we get up to? What problems are such perturbations causing? What, if any, important services do these biogeochemical cycles provide to us?

We two authors have devoted ourselves to understanding the silica cycle. We think it's great and would love to explain it ALL to you. But we'll aim instead for not going too far overboard and scaring you away before the last chapter, which is the most important and (we think) most interesting one.

If you put a gun to our heads and said (or, really, just asked us even half-way nicely), "Break down the silica cycle (as it occurs today) into six simple steps!" this is what we'd tell you.

One: when the silicate minerals of continental crust weather, they release dissolved silica to soil water, ground water, or river water. When oceanic crust weathers, dissolved silica is released directly to the sea.

Two: although a lot of the silica liberated by continental weathering becomes trapped in secondary minerals such as clays, a portion of it makes its way to the ocean, although not before cycling through the terrestrial biosphere numerous times.

Three: today, dissolved silica in the ocean gets turned into solid silica by diatoms, radiolarians, sponges, choanoflagellates, silicoflagellates, and other silica-biomineralizing organisms.

Four: solid silica settling in the sediments gets buried by material settling on top of it.

Five: as part of the creation, migration, and destruction of ocean floor via plate tectonics, eventually the sedimented silica gets dragged down into the interior of the Earth to join the super-hot viscous mantle.

Six: mantle intruding upwards into either continental or oceanic crust (volcanically, for instance) cools to form solid silicate minerals which, in a return to step one, may now be weathered to release dissolved silica.

And now we're back at step one.

Every biogeochemical cycle deserves a picture, so, although we're more scientifically than artistically inclined, we've drawn you one (Fig. 9.1). Isn't it fabulous?[1] It shows the whole big picture.

The weathering of newly produced continental silicate minerals begins in the upper left corner of the diagram. From there, silica cycles through the terrestrial biosphere on its way to the ocean. At the same time, in the middle of the picture, dissolved silica is added directly to seawater through the weathering of oceanic crust at mid-ocean ridges. Up at the top of the ocean, dissolved silica taken up by silica biomineralizers like diatoms becomes incorporated into biogenic silica, most of which dissolves before it manages to sink all the way to the seafloor. The solid silica which survives to become buried in seafloor sediments rides along as oceanic crust (and lithosphere) gets pushed from the mid-ocean ridge to the ocean margin to be subducted back down into the mantle to someday be reborn again as new continental or oceanic crust.

---

[1]Yes, we know—*Don't quit the day job.*

9.1 The Silica Cycle

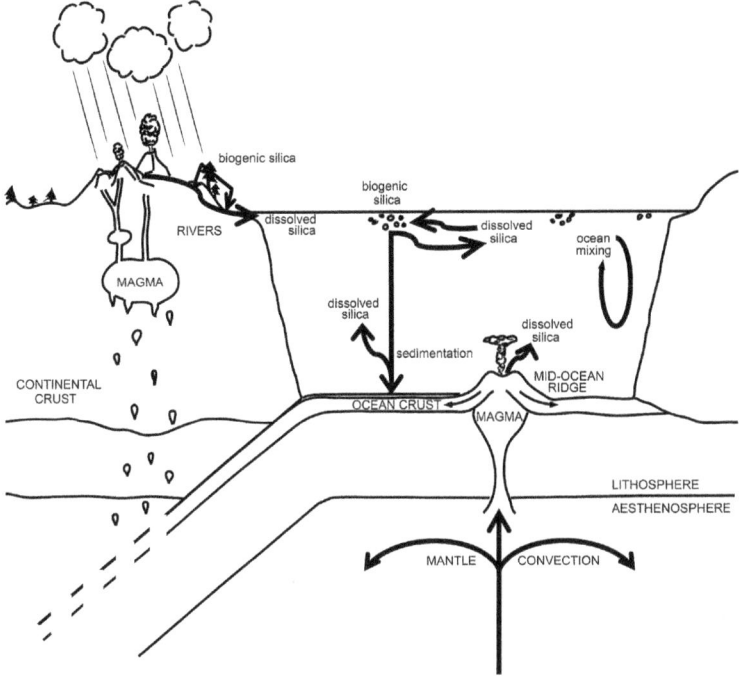

Fig. 9.1 The modern day silica cycle

We're going to take the rest of this chapter to repeat all of that yet again, but in greater detail. It helps to have the general layout of the forest in your head before zooming in and having a gander at the different trees.

9.2 Silicate Weathering

Right here, right now, we're here to talk to you about silicate weathering. This is the one step in the silica cycle that everything ultimately comes back to. And by everything we mean, with only slight exaggeration, *everything*.

Never mind that most of the silica we've talked about so far in this book—that which has found its way into the likes of rivers, lakes, streams, plants, chickens, sponges, people, microorganisms, sewage, and beer—has ultimately (or even proximally) come from the weathering of primary silicate minerals,[2] your life literally depends on silicate weathering. Silicate weathering is, in concert with the

---

[2]Primary silicate minerals are the minerals that formed during the initial solidification of the molten silicate from the interior of the Earth. In contrast, secondary minerals are those which form due to the chemical or mechanical alteration of primary minerals.

Sun, the one main thing that keeps our planet habitable. As the only inhabitants of this planet intelligent enough to understand it and as the potential terraformers of a few planets nearby, we ought to all feel a duty to understand silicate weathering, at least at a basic level.

The basics of silicate weathering, although simple enough to understand, are kind of mind-blowing. Fully appreciating silicate weathering is no less satisfying, although it can take decades and still there will be aspects of weathering that remain elusive. It is something that happens within a system of complex, interlocking biogeochemical cycles with positive and negative feedback loops galore, interplay with climate, and increasing influence from the biosphere.

To begin to understand silicate weathering, it helps to identify the chemical reactions that are involved. The easiest way to do this is to write out chemical equations based on the basic rules of inorganic chemistry and a knowledge of what primary and secondary minerals are reacting and being produced. That may sound awful, but high school chemistry is all you need to follow along. The best bet, no matter how much chemistry you've had, is to start with the simple, such as with the weathering of the mineral wollastonite.

Wollastonite[3] is a calcium silicate mineral whose chemical composition is $CaSiO_3$. That's as minimalist as silicate minerals get (outside of a plain old pure silica like quartz). The calcium atoms in wollastonite have taken the place of one oxygen atom in each silica tetrahedron and their presence stabilizes the helical rows of silica tetrahedra that make up the mineral.

Because wollastonite doesn't have all sorts of other ions, like aluminum or potassium, nosing in and complicating things, the weathering of wollastonite is the first (and sometimes only) weathering reaction a student of geochemistry learns. It goes like this:

$$CaSiO_3 + 2(H_2O) + CO_2 \rightarrow Ca^{2+} + Si(OH)_4 + CO_3^- \qquad (9.1)$$

What this balanced equation reveals is that for each molecule of dissolved silica ($Si(OH)_4$) weathered from wollastonite, one calcium ion (Ca^{2+}) is also liberated. Thus ultimately the weathering of silicate minerals is not only where all the dissolved silica in river and ocean waters has come from, it is also where all the dissolved calcium and all the other dissolved salts have come from.[4]

Most interestingly, producing dissolved silica ($Si(OH)_4$) from wollastonite requires protons (aka hydrogen ions, H^+). These protons come from water because

---

[3]We were almost sad to discover that wollastonite was not named for Roland Wollast, one of the great recent heroes of biogeochemical research, but for William Hyde Wollaston (1776–1828), a wide-ranging scientific genius whose insights, inventions, and discoveries in just one of the many fruitful areas of his work laid the groundwork for the modern field of crystallography.

[4]Although proximally most of the dissolved calcium has come from the dissolution of secondary calcium carbonate minerals formed from calcium that ultimately came from primary silicate minerals.

9.2 Silicate Weathering

the mineral itself doesn't have any. But for water to dissociate[5] enough to contain a usefully ample supply of H^+ ions requires acidity. That's where the carbon dioxide (CO_2), dissolved into the water from the air, comes in. Carbon dioxide is a great acidifier. Without it much less chemical weathering would occur.

So now this is *really* interesting. The dissolution of wollastonite doesn't just produce calcium ions and dissolved silica, it also converts one molecule of a powerful greenhouse gas into a bit of what we call alkalinity. In this case, that alkalinity is a carbonate ion (CO_3^-) that will probably be someday used to construct the calcium carbonate ($CaCO_3$) of a calcitic or aragonitic seashell or coral.

The flashing neon sign here is: *the dissolution of silicate minerals consumes carbon dioxide*.

Hold on to that thought. It's the key to the long-term habitability of Earth and we'll get back to it in the last chapter of this book, the one where silica rides in and saves the day. But right now we can't stop here (even if most students of geochemistry do). The weathering of wollastonite is for *amateurs*.

Wollastonite weathering's simplicity makes it a great thought experiment and a good sign that silicate weathering (and the silica cycle) has ties to carbon dioxide. But in real life, hardly any silicate minerals are wollastonite and as a model of weathering, its limits are severe.

Most silicate minerals contain, in addition to the requisite silica tetrahedra and the optional but common calcium atoms, elements like aluminum, potassium, sodium, fluorine, iron, zinc, manganese, magnesium, boron, lithium, titanium, and/or sulfur, meaning that there are lots of very different ions unleashed during silicate weathering. That makes for some seriously cluttered up weathering equations. Worse, if weathering releases enough solutes for the local waters to become saturated relative to some mineral, generally a clay, that (secondary) mineral will precipitate. This precipitation will remove some of the silica and some aluminum and/or other ions from the water. In other words, wollastonite weathering led us astray. The process of chemical weathering is part dissolution *and* part precipitation.

A (more) realistic weathering equation looks like this one for the dissolution of the primary potassium aluminosilicate mineral orthoclase ($KAlSi_3O_8$) (which is a solid) to produce dissolved silica ($Si(OH)_4$), potassium ions (K^+), and bicarbonate ions (HCO_3^-) (which are solutes), and the clay mineral kaolinite ($Al_2Si_2O_5(OH)_4$) (which is a solid):

$$2(KAlSi_3O_8) + 11(H_2O) + 2(CO_2) \rightarrow Al_2Si_2O_5(OH)_4 + 4(Si(OH)_4) + 2(K^+) + 2(HCO_3^-) \quad (9.2)$$

That's a mess, but kind of a hot mess. Although the weathering of orthoclase is more complicated than the weathering of wollastonite, at its core the base truth remains true: the weathering of this silicate mineral has released silica to solution,

---

[5]The dissociation reaction is: $H_2O \leftrightarrow H^+ + OH^-$

freeing it for active participation in the cycling of silica through the marine and terrestrial biospheres, and it has converted carbon dioxide into a bit of carbonate alkalinity (in this case the ion bicarbonate).

But for all its messiness, the orthoclase weathering equation is also oversimplified and idealized. Chemical weathering reactions are written in the manner of all chemical reaction equations. They are written as if they are occurring in isolation and as if they are going all the way to completion.

But in the real world those two conditions are rarely met. Solutes can be whisked away in moving water before they reach concentrations high enough for precipitation of a secondary mineral, preventing the reaction from going to completion. Also because silicate rocks are generally made of a consortium of silicate minerals, generally many different silicate minerals are dissolving at the same place and time. So lots of different weathering reactions may be happening together, releasing all sorts of different solutes in all sorts of differing quantities. Who is to say (besides a computer model designed to keep track of all the reactions that could occur) which and how many secondary minerals will form and when. And don't even let's get started about uptake of solutes into the biosphere; that's even harder to predict.

But even in the more complicated, complex mish-mash of circumstances of the real world, the core truth of the chemical weathering of silicate rocks remains the core truth: in addition to releasing dissolved silica (and other materials) to solution, silicate weathering converts carbon dioxide into alkalinity in the form of carbonate ions and/or bicarbonate ions.

That's a pretty significant core truth. Carbon dioxide is a greenhouse gas that heats up the surface of the Earth. But carbonate and bicarbonate ions are solutes better known for settling upset stomachs and making it possible to be a clam.

9.3 Getting Silica from Continent to Ocean

After silicate weathering has produced dissolved silica, if that weathering has occurred on a continent, the next task for that liberated silica is to get to the ocean. But as you, having read Chaps. 5 through 7 about silica biomineralization, animals' dietary need for silica, and the production of silica phytoliths by most land plants, getting to the sea generally involves detours.

Not all of the detours are temporary. Some silica fails to reach the ocean. It has gotten trapped into secondary minerals such as clays. Although some of these clays will dissolve at some point, rereleasing silica to participate in the silica cycle, many don't. Of those that don't dissolve, some will accumulate upon the land, becoming buried, never again to see the light of day. Most clays, however, even if it literally take eons, will wash out to the sea. There they slowly settle and become buried in ocean sediments that eventually become subducted down and subsumed into the mantle to be recycled back out as new silicate rock.

9.3 Getting Silica from Continent to Ocean

What silica doesn't get trapped into secondary minerals leaves the soil or other site of weathering in a reactive phase[6] (such as dissolved silica or some sort of amorphous (noncrystalline) solid silica) and makes its way into rivers that head toward the sea. This is represented by the arrow moving from mountain to ocean in the upper left corner of Fig. 9.1.

In this day and age of farms, fields, forests, jungles, grasslands, and etc, reactive silica doesn't have a straight shot to the sea. It tends to cycle along the way numerous times through the terrestrial biosphere. Sometimes dissolved silica gets drunk and finds itself promoting the production of collagen in an animal (as in Chap. 6). But more likely (or rather, more likely first), given that your typical modern day plant is somewhere between one and ten percent silica by dry weight and that Earth has no shortage of land plants, dissolved silica is taken up into plants and precipitated as opal phytoliths. As detailed in Chap. 7, these particles of plant silica eventually drop back into soils directly or via an herbivorous or omnivorous gastrointestinal tract (or they end up in the maze of somewhat unknown final destination that is sewage systems). Dissolved silica can also be taken up by freshwater diatoms, freshwater sponges, or siliceous soil microbes and fixed into what biogenic silica they produce (as discussed in Chap. 5).

Never mind what land animals and freshwater diatoms and sponges are doing with silica (which is a lot), every year land plants are producing silica phytoliths to the tune of somewhere between 60 and 200 teramoles of silica (aka 60×10^{12}–200×10^{12} mol of silica, or 3.6×10^{37}–1.2×10^{38} molecules of silica, a mole being 6×10^{23} items of something, such as atoms or molecules). That is 10 to 33 times the amount of silica we think is being liberated each year by silicate weathering on the continents. And that indicates that the biological cycling of silica on land is highly dynamic. For plants to produce 10–33 times more silica than is coming in from outside the terrestrial biosphere requires that, once caught up inside the terrestrial biosphere, each silica tetrahedra is taken up, incorporated into phytolith silica, and then dissolved from biogenic silica one to three dozen times before its escape to the ocean. That's a lot of looping that, again, doesn't count what animals, freshwater diatoms, and freshwater sponges are getting up to with the silica.

While some tiny portion of the terrestrial biogenic silica that is produced each year will become trapped in soils, lake sediments, or, these days, in the sediments upstream of any one of the Earth's 850,000 man-made dams (as explained in Chap. 8). However, the ultimate fate of most of the silica that cycles through the terrestrial biosphere is to reach the sea.

Today, rivers and the release of groundwater through submarine springs deliver 85% of the reactive silica that enters the oceans. Our best current estimate of the

---

[6]Meaning it is either already dissolved silica, which can participate in all sorts of chemical reactions and be taken up biologically, or it is in the form of some easily dissolvable solid silica, generally a noncrystalline variety such as biogenic silica or a non-biogenic amorphous silica.

quantity is 6 teramoles of silica (which is 3.6×10^{36} molecules of silica). In a perfect, balanced world without dams, eutrophication, and a relatively recent change of much of the land surface from wilderness to farm, pasture, road, and city, 6 teramoles would also be the net amount of silica weathered out of continental rock each year.[7]

Although a number of silica scientists have derived the 6 teramole number (somewhat) independently via several different approaches (measurements, modeling, and mixtures of both), 6 teramoles of silica is a number that we know only roughly. The correct number could easily be higher or lower by 20% (in other words, it could be anywhere between about 5 and 7 teramoles). That is a large margin of error. It's like saying you know that you're somewhere between 50 and 70 years old so you've settled on calling yourself 60.

Part of the reason for the large amount of uncertainty is, well, let's put it this way—have you ever tried to make a global estimate of river input of anything into the ocean? It's not an easy thing to do. There are tens of thousands of rivers and to obtain a good estimate of the input you can't just sample a good portion of them just once. You'd need to measure their water flow and their dissolved silica concentration frequently throughout the year for a set of (not necessarily consecutive) years perhaps for decades to avoid getting a bad answer due to short-term variability in climate and rainfall. And another thing: for a complete estimate, you'd also have to monitor groundwater discharge into the coastal ocean via submarine springs all over the world.

Does that sound hopeless so why bother? But science isn't about perfect answers, it's about working toward the most complete, accurate, and precise answer possible and then working on improving it. Science is about figuring out how systems function (and how the world works) while trying not to lose sight of the limitations of the answers at hand. In terms of estimating how much reactive silica reaches the ocean each year, as the decades roll by, the number of measurements hundreds of silica scientists and other biogeochemists have made all over the world adds up. At the same time, the computer modeling we're working on to make sense of the data grows more sophisticated and set into better geographic perspective. So our estimate for the input of reactive silica to the oceans is getting better.

Sometimes better means more accurate (closer to the true value). Sometimes it means more precise (with less uncertainty associated with it). Sometimes it means we have a better idea of the shortcomings of the estimate. And sometimes it means we've been able to identify more sources, sinks, and processes that we need to keep track of to get the full picture of what's going on.

---

[7] Net means that the number doesn't include the amount of silica trapped permanently in secondary minerals such as clays, a quantity that could be up to two-thirds of the gross amount weathered out of primary silicate minerals.

9.4 The Weathering of Oceanic Crust

Let's hop now to the bottom of the ocean, because if 85% of reactive silica comes into the ocean via rivers and groundwater, 15% has to come in through other means. Most of those other means are deep sea hydrothermal vents and warm springs associated with the creation of new oceanic crust. This input is represented at the bottom of the ocean in the silica cycle sketch (Fig. 9.1) because where you find ocean crust is underneath the sediments that have piled up on the ocean floor.

We know from measurements made using such modern technologies such as sonar, magnetometry, and seismic tomography, that below both the continental and the oceanic crust there is mantle, that silicate 84% of Earth that surrounds the nickel–iron core of the planet. The mantle itself consists of two slightly different portions, an upper portion and a lower, more interior portion. These are marked on Fig. 9.1 as lithosphere (the shallower portion) and aesthenosphere (the deeper portion).

The aesthenosphere, being deeper within the Earth and therefore some distance from the crust through which mantle heat is lost, is the hotter portion of the mantle, reaching temperatures of 4000 °C (7200 °F). By all rights, the solid silicate that makes up the aesthenosphere should be liquid. But the extreme lithostatic pressure of the deep Earth keeps this stuff solid despite the heat. (Extreme lithostatic pressure, by the way, is a fancy way of saying that the pressure in the aesthenosphere is high because there is an awful lot of planet piled on top of it.) But this does not mean the silicate rock of the aesthenosphere is as solid as the silicate rock of the Earth's surface. The hot solid silicate of the aesthenosphere is ductile enough to slowly flow, somewhat like glaciers and ice sheets do even though they are also solids.

This slow flow-ability makes gigantic and gigantically long-lived convection cells possible within the aesthenosphere. And so what is happening right now (and has been happening for billions of years) is that super-hot material from deep in the aesthenosphere, hot things generally being more buoyant than colder things, rises up to the Earth's crust in some locations. To balance things out, cooler mantle material sinks back down toward the Earth's core at other locations.

You can see the top of two such convection cells at the bottom of Fig. 9.1, where, in this two-dimensional rendering, upward moving mantle hits the base of the lithosphere and goes three ways. Two of those ways are along the base of the lithosphere until the material cools enough to become dense enough to sink back down into the deep aesthenosphere, completing the convection cell (or, rather, two of them).

The third way is that some of the heat and hot silicate material pushes all the way up through the lithosphere to emerge out onto the ocean floor. Because the pressure at the bottom of the ocean is much lower than the pressure within the aesthenosphere, what mantle emerges is initially not just ductile but molten. But the cold,

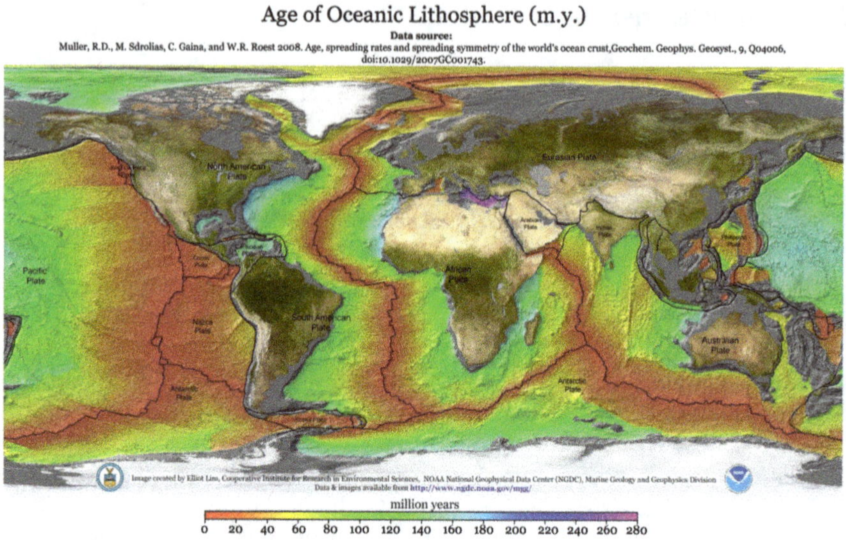

Age of Oceanic Lithosphere (m.y.)

Data source:
Muller, R.D., M. Sdrolias, C. Gaina, and W.R. Roest 2008. Age, spreading rates and spreading symmetry of the world's ocean crust, Geochem. Geophys. Geosyst., 9, Q04006, doi:10.1029/2007GC001743.

Fig. 9.2 Map showing the age of the seafloor and the location of mid-ocean ridges. This image was created by Mr. Elliot Lim, CIRES & NOAA/NCEI, with data from Müller et al. (2008) Geochem Geophys Geosys 9, Q04006, doi:10.1029/2007GC001743 and has been reprinted with permission. The original full color version of this image and many other excellent images of the age of the ocean floor can be found at https://www.ngdc.noaa.gov/mgg/image/crustalimages.html

deep waters of the ocean cool it quickly and it solidifies and then Bob's your uncle, as the Brits would say, brand new oceanic crust.

This is what happens along the central axis of mid-ocean ridges. On Fig. 9.2, you can see where these mid-ocean ridges snake through the oceans: longitudinally down the middle of the Atlantic Ocean, for example, but up the eastern side of the Pacific, until about 9°N.

Plate tectonics, that shoving of the continents, happens because brand new ocean crust doesn't remain the brand newest ocean crust for long. More molten mantle quickly follows, bisecting the previously newest bit of ocean crust and shoving half to one side of the ridge and half to the other. Thus are the oceans spread apart a centimeter or two per year over tens to hundreds of millions of years.

Because a mid-ocean ridge creates material equally along both sides of its axis, all mid-ocean ridges should run down the center of their ocean basin. However, things can go asymmetric when geological processes such as subduction[8] occur more strongly on one side of an ocean basin than on the other. This can happen

---

[8]During subduction, the slab of oceanic crust, the oceanic lithosphere just beneath it, and the sediments that have collected on top of it, having met an obstacle such as a continent, gets pushed down and begins to descend into the mantle, as show in Fig. 9.1. By the way, because oceanic crust is richer in magnesium and iron and poorer in aluminum than continental crust, it is the denser of the two. When the twain collide, the lighter ends up on top and the heavier is forced down into the mantle.

9.4 The Weathering of Oceanic Crust

when, for example, a descending plate of lithosphere gets jammed in the subduction zone, halting further subduction. (This is exactly what happened to the subduction zone that should be offshore of California and Oregon but isn't (have a close look at Fig. 9.2). Thus one of the Pacific Ocean's major spreading ridges, the East Pacific Rise, very much so does not run down the center of the Pacific Ocean basin.)

Here now is where we finally get back to the silica cycle and that remaining 15% of the reactive silica that gets added to the ocean each year.

The mantle rising upwards under mid-ocean ridges does not merely open up ocean basins and shove continents around, it also results in the weathering of fairly young oceanic crust. This also happens because heat rises. In this case what is created is the convection of water through the oceanic crust.

This hydrothermal weathering of oceanic crust begins when new crust cools and then, as newer crust forms, gets shoved away from the ridge center. This causes the young crust to crack and fissure. Cold seawater sinks into these openings. As it draws closer to the mid-ocean ridge's magma chamber, it begins to heat up. As it warms, it begins to weather the oceanic crust.

This hot seawater swaps calcium with the young crust in an almost even exchange. It also leaches metals like iron and manganese out of the crust until the waters are saturated with them. The hot water also loses dissolved oxygen until there is none left but gains a lot of dissolved carbon dioxide.

And, of course, the hydrothermal weathering of the ocean's silicate crust adds a lot of dissolved silica to the fluid. By the time this hydrothermal fluid née cold ocean water has become hot enough to rise and vent dramatically back out to the ocean, it is fully saturated with dissolved silica. This dissolved silica mixes itself into the cold, undersaturated deeper waters of the ocean, as implied by the venting out of the mid-ocean ridge on the silica cycle sketch (Fig. 9.1).

Since the discovery of deep sea hydrothermal vents back in the 1970s, hundreds of scientists have studied them using submersibles, remotely operated vehicles, benthic landers, video cameras, all sorts of ingenious sampling techniques and equipment, and naturally occurring geochemical tracers. So we have a reasonable estimate of the amount of dissolved silica that is venting out of high temperature hydrothermal vents of mid-ocean ridges.[9] When we add the amount of reactive silica added to the oceans via deep sea hydrothermalism to the amount of reactive silica delivered by river and ground water, we get a value of approximately 9 teramoles of silica (give or take 50%). That's how much "new" silica is being added to the oceans every year due to the weathering of primary silicate minerals.

---

[9]How much is coming in or being lost through lower temperature, more diffuse flow through ridge flanks, however, is a serious issue that remains to be solved.

9.5 Silica Biomineralization in the Ocean

Today, once added to the ocean, dissolved silica is available for use by silica biomineralizers such as diatoms, radiolarians, silicoflagellates, choanoflagellates, and sponges. Deep sea sponges aside, most of this rogues gallery of silicifying organisms lives in the sunlit surface layer of the ocean.[10] Furthermore, because our friends the diatoms are impressively numerous, fast-growing, and notably siliceous, it is a safe bet that most of the 240 teramoles (240×10^{12} mol aka 1.4×10^{10} metric tons) of biogenic silica produced in the upper ocean each year is being produced by diatoms. Thus the production of biogenic silica in the oceans is depicted in the upper part of the ocean on the silica cycle sketch (Fig. 9.1). The diatoms are not to scale.

The fate of almost all of this biogenic silica that is made each year is to rapidly dissolve. The modern day ocean is after all extremely undersaturated with respect to noncrystalline silica. For this biogenic silica not to dissolve, concentrations of dissolved silica in ocean waters would have to be close to 1000 micromoles per liter rather than today's 0–15 micromoles of dissolved silica per liter in most surface waters and, with rare exception, less than 200 micromoles of silica per liter in even the most siliceous deep waters, the ones in the far northern Pacific.

So strong is the power of this undersaturation, slightly more than half of the biogenic silica produced each year dissolves even before it has had time to sink only 100 to 200 meters. A further portion of each year's biogenic silica crop dissolves while sinking through the cold, dark, deep interior of the ocean, and yet more dissolves while biogenic silica sits on the seafloor waiting to be buried by particles settling down on top of them. In the end, only 2–3% of the biogenic silica produced in the oceans each year becomes permanently buried in ocean sediments.

Efficient this biological pumping of silica out of the ocean is not.

But permanent export of 2–3% of each year's crop of biogenic silica is enough to (more or less) equal the amount of reactive silica coming in to the ocean via rivers, submarine groundwater springs, and mid-ocean ridge hydrothermal fluids. And because the gross amount of biogenic silica production is so high, a removal efficiency of 2–3% is enough to keep ocean waters all but entirely depleted of dissolved silica.

---

[10]To put it somewhat bluntly, the critters are either photosynthetic or eating things that are.

9.6 Silica's Return to the Mantle

Silica may have exited the ocean by getting itself buried in deep sea sediments, but its cycle is not yet over. Deep sea sediments collect on top of oceanic crust. The silica within the sediments therefore rides the plate tectonic railway away from the nearest mid-ocean ridge. But this is more a slow, pokey freight train (albeit one with all but unstoppable momentum) than it is the transoceanic express. Ocean floor is typically moving only one to two centimeters per year away from the mid-ocean ridge that generated it. That's less than an inch.

One of the results of ocean crust being created at ridges, split in two, and then pushed away to both sides is that if you're looking for brand new oceanic crust, you will find it at the central part of a mid-ocean ridge and if you're looking for the oldest oceanic crust, you need to look for it at the edges of the ocean basin.

Some of the oldest oceanic crust you can find, clocking in at roundabouts 180 million years, abuts the eastern coast of the United States. Its counterpart abuts northwest Africa. Both can be seen in Fig. 9.2, which is a map of the age of the ocean floor all over the world. There aren't subduction zones along these coasts—thus ocean crust isn't being destroyed here—and so 180 million years ago is when the North Atlantic ocean basin began rifting open in the midst of an enormous land mass. This creation of the North Atlantic was part of the breakup of the supercontinent Pangea (which is pictured the upper right panel of Fig. 9.3 in the form it had 195 million years ago) into chunks that included what we now call North America, Africa, and Eurasia.

But almost all of the oceanic crust that exists today is much younger than 180 million years. Oceanic crust normally finds itself in a subduction zone more than 100 million years earlier than that; the current average life expectancy of oceanic crust is only 70 million years. In case you're wondering, 70 million years is less than 2% of the 4 billion year age of the Earth.

Oceanic crust has such a geologically meager life expectancy because the Earth can't keep making new crust without something giving. Otherwise by now the surface of the Earth would be stacked floor to ceiling with oceanic crust and the interior of the Earth would be missing prodigious quantities of mantle. Physically, it isn't possible. Thus for every ocean basin widening, another is closing down.

Today, for example, the continued opening of the Atlantic Ocean basin is accommodated by the shrinking of the Pacific Ocean basin. It isn't that the Pacific isn't producing new oceanic crust at its mid-ocean ridges, but oceanic crust is being subducted down into the mantle along its rim faster than it is being produced at its mid-ocean ridges. The spreading ocean crust in the Pacific is pushing Asia and the Americas apart against the force of the Atlantic pushing the Americas away from Europe and Africa, and the Atlantic is winning. Thus, with some interesting but now reasonably well understood exceptions in some regions (such as along the coast of California), where the oceanic lithosphere of the Pacific meets the lithosphere of the continents surrounding it, subduction is occurring.

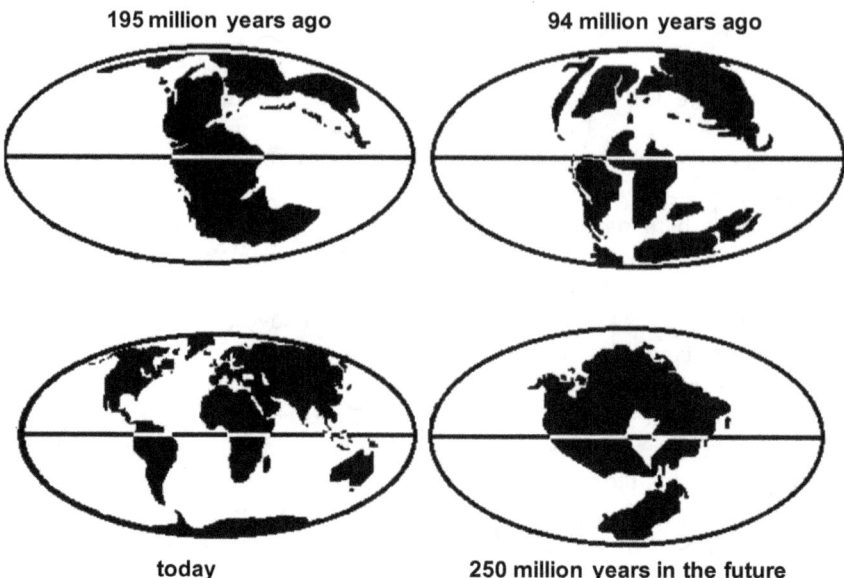

Fig. 9.3 The position of the continents 195 million years ago, 94 million years ago, today, and 250 million years in the future. These images are based on those of the plate tectonic maps and continental drift animations by CR Scotese, PALEOMAP Project (www.scotese.com) and included here with permission

Over the eons, this plate tectonics which closes the circuit of the long-term silica cycle, gets stuff done. The continents are pushed together and apart, crashed into each other with startling result (such as the Himalayas), and forced to wander hither and yon across the surface of the Earth. Nearly two centuries of work by field geologists, geomagneticists, geophysicists, and even paleontologists has gone into figuring out these dances. And they are amazing.

Figure 9.3 shows some screen shots, as it were, of the wanderings of Earth's continental plates over the last 200 million years and of what is expected to happen over the next 250 million years due to the continued production and subduction of oceanic lithosphere. Sometimes all the land is huddled together in one massive mass known as a supercontinent. Other times the land has been broken apart into continents far apart from each other all over the Earth. (Either way, the totally continental area over the last several hundred million years has remained more or less the same.)

To get back to completing the circuit that is the silica cycle, as we've mentioned before, as a slab of oceanic crust moves down into the mantle, it takes most of its sediments with it.[11] As the crust and its sediments descend together through the

---

[11]Most instead of all because some of the sediments get scraped off against the continental mass and accumulate at that intersection (or even get pushed up onto the continent itself).

9.6 Silica's Return to the Mantle

upper mantle toward the aesthenosphere and temperatures rise, the biogenic silica in the sediments melts. At the same time, any calcium carbonate there melts, too, and organic matter becomes oxidized. This returns the silica (and the calcium, and the carbon, and whatever other elements are present) back to the mantle.

Because subduction occurs where continent meets ocean and the diving plate of oceanic lithosphere is moving toward the continent, the melting of the oceanic lithosphere and its sediments occurs underneath the continent, not too far from the continent's edge (see Fig. 9.1). When the resulting melt rises, it can erupt as silicate lava out of a volcano or solidify as the igneous batholith at the core of a mountain range. That's how the western United States got the Sierra Nevadas and South America got the Andes, for example. And this then bring things finally full loop to new continental crust made of silicate minerals.

In the long run, the balance is close to perfect. Descending oceanic lithosphere returns to the mantle silicate minerals produced from mantle material at mid-ocean ridges. The melting of the sediments riding along with this descending slab of lithosphere to produce new continental silicate minerals replaces the continental silicates that were chemically weathered at the beginning of the silica cycle.

To a biogeochemist with cycles on the brain, it doesn't get much more satisfying than that.

9.7 The Earth's Early Ocean Was a Tremendously Siliceous Place

One of the most interesting things about biogeochemical cycles is that they have pretty much all changed drastically over Earth's history. This is super true of the silica cycle. In the silica cycle as it operates today, silica weathered out of silicate minerals on the continents and seafloor, cycles through the terrestrial and marine biospheres on its way toward burial as biogenic silica in marine sediments that are ultimately subducted back down into the mantle. But in the beginning and for 1.5–2.5 billion years afterwards, Earth's silica cycle operated without any intentional biological production of silica at all.

On the continents, there was no dynamic cycling of silica into and out of land plants, leaf litter, soils, diatoms, lake sediments, freshwater sponges, and collagen-producing animals. Meanwhile, the oceans held no diatoms, sponges, or anything capable of precipitating silica under conditions of extreme undersaturation. The permanent removal of silica from the oceans so it could be recycled down into the mantle by plate tectonics could only have occurred through the abiotic precipitation of silica particles and cements, almost all of which require excessively high concentrations of dissolved silica in the ocean water to occur. Which means that, at least in terms of dissolved silica, for most of its existence, the great ocean that belongs to Earth was a totally different environment than it is today.

In addition to lacking silica biomineralizers, the Earth during this ancient Archean period was still hot-blooded from its formation. Hydrothermalism would have been at an all-time high, spewing dissolved silica into the ocean at an incredible rate. One important abiotic mechanism for removing dissolved silica even at undersaturated concentrations, the adsorption of silica onto the surfaces of clays and zeolite minerals, limited by chemical kinetics and by the restricted amount of those minerals exposed to seawater, would have been unable to keep pace with the inputs of silica. This would have guaranteed that dissolved silica concentrations in the ocean then weren't just high but saturated and not just saturated but supersaturated. By some point in the Early Archean, concentrations of dissolved silica in the ocean would have reached the point where amorphous silica could precipitate out of the ocean on its own. We can put a number to that. That number is more than 4000 micromoles of dissolved silica per liter of seawater.

That is a lot of dissolved silica.

There are two different types of geologic evidence to back up that theory. The first comes in the form of the trace elements and of the ratios of the stable isotopes of silicon contained within marine sediments that were silicified 3 to 3.5 billion years ago and are now found in the Pilbara Craton of Western Australia. These chemical fingerprints can reveal the source of the silica in the cements—was it local, hydrothermal source, or was it the ocean itself—because two types of fluid contain silicon of different stable isotopic composition and different ratios and amounts of trace elements. From these chemical signatures, we can see that the silica that cemented these ancient sediments together did not come from a hydrothermal source, but had to have precipitated directly from Early Archaean seawater as a whole. The only way that could have happened was if Early Archean seawater had indeed been excessively supersaturated with dissolved silica.

The second line of evidence is way cooler. If you microscopically inspect some of the 3.5–3.2-billion-year-old rocks from South Africa's Barberton Greenstone Belt and from the Pilbara Craton you'll find that it's full of miniscule granules of what is now almost perfectly pure microcrystalline quartz. Some of these granules, which started out life as amorphous silica but (inevitably) crystallized into quartz as the eons trudged by, are pleasantly spherical. Others are cutely oblong, as if slightly squashed. All have been cemented into place by silica that precipitated around them, which means that they probably didn't initially form within the sediment. The most likely scenario is that they precipitated out of seawater somewhere high above the seafloor and then sank down. This would explain the oblong ones as victims of later silica blobs that land on top of them.

Sometimes we just kick back and imagine that. Earth's early ocean was so siliceous, from time to time and place to place during the Early Archean there were underwater rainfalls of silica droplets that fell upon the ocean floor and then became entombed in a self-forming silica glue.

9.8 Silica, Cyanobacteria, and Banded Iron Formations

Eventually the raging of Earth's hydrothermalism slowed down some, diminishing the hydrothermal input of silica into the oceans. Adsorption of silica onto mineral surfaces was able to keep up and even to bring dissolved silica concentrations down somewhat below supersaturated. Never again (or at least not so far) would silica be spontaneously precipitating in the middle of the water column of the deep blue sea.

But there was still enough dissolved silica and, because the oceans and atmosphere were still lacking in molecular oxygen (O_2), there was also enough dissolved iron for these things known as Banded Iron Formations (BIFs) to form.

The BIF are extremely ancient rock formations that consist of alternating bands of black and red silica, which is how they got their name. The black bands consist of an oxidized iron mineral (such as magnetite, Fe_3O_4, or hematite, Fe_2O_3) cemented in place by silica. The red bands consist of a more pure silica and contain very little iron. BIFs began forming perhaps as early as 3.7 billion years ago, reached the peak of their formation, some 2.5 billion years ago, and then 1.9 billion years ago, their production entirely ceased. Although the details of how and when the iron got oxidized to form the BIFs, cyanobacteria are commonly implicated and almost certainly critical, especially to the formation of BIFs that are 2.5 million years old and younger. Between the chemical composition of their cell membrane to the chemistry of their photosynthetic activities, cyanobacterial are good at causing iron oxides to precipitate. The iron oxides in turn are excellent at scavenging silica from solution, causing it to form solid, amorphous silica. This may mean that cyanobacteria were the first biological participants in the silica cycle.

Do you know these blue-green algae? Maybe you've experienced the joy that is degusting the deeply green nutritional supplement known as *Spirulina*.[12] But the original claim to fame of cyanobacteria (or maybe of the not-quite-cyanobacterial last common ancestor of all cyanobacteria) was as the first organism on Earth capable of oxygenic photosynthesis.

Yep, it is possible to photosynthesize without producing O_2. Instead of splitting oxygen-liberating H_2O to obtain electrons and hydrogen ions (H^+), anoxygenic photosynthesizers use an oxygen-free compound such as hydrogen sulfide (H_2S).

This anoxygenic photosynthesis was the first type of photosynthesis to be developed here on Earth and it ruled the roost for a long time, too—about two billion years. But it's sort of a 1.0 version of photosynthesis, with plenty of room for improvement. For instance, a photosynthesizer is far more likely to run out of H_2S than water, there being literally oceans of the latter on Earth but not of the former. Possibly as early as 3 billion years ago but most certainly before 2.4 billion years ago, the last common ancestor of all currently existing cyanobacteria hit the evolutionary jackpot: the molecular machinery, acquired over tens to hundreds of

---

[12]People have been cultivating and consuming *Spirulina* (now more correctly known as *Arthrospira*), a genus of freshwater cyanobacteria, since at least Aztec times. Which goes to show how desperate people can get when they're hungry.

billions of generations, capable of breaking up water during photosynthesis. One pretty major side effect of this, by the way, was the oxygenation of Earth's atmosphere and oceans.

But to reach anything like today's nearly 30% oxygen atmosphere took some time. So at first cyanobacteria were great for BIFs because they were good at oxidizing iron.

Many arguments (not all relying upon cyanobacteria) have been put forth to explain the formation of BIFs. Although for different times and places in early Earth history, different ones of these arguments will be closer to the truth than others, this one is our favorite. Cyanobacteria can be benthic, growing in films upon surfaces (such as the seafloor) in water shallow enough to be sunlit. Cyanobacteria living on the seafloor in shallow, subtidal areas of the ancient ocean drove the precipitation of iron oxides directly on the seafloor itself. These iron oxides that were already on the seafloor then collected silica. But cyanobacteria can also be planktonic, growing freely in the water itself. Planktonic cyanobacteria would have been producing particles of iron oxides up in the water column. These iron oxide particles, which would have immediately started to collect silica, would have then settled down onto the shallow seafloor.

In both of these cases, during sunny spring, summer, and early autumn (good times for cyanobacteria to be photosynthesizing) cyanobacteria would have been producing iron oxides that ended up at the bottom of the ocean (although only in shallow water), collecting silica all the while. This is the black band of a BIF. Come winter, however, sunlight would have been limiting to photosynthesis and rates of iron oxide formation would have diminished. But silica would have continued to collect, growing up from the nucleation surface that the iron oxide initially provided. The addition of solid silica would have continued steadily on through the winter, resulting in the red, relatively iron-poor band of the BIF. Come spring sunshine again, though, and the next layer to go down would be deeply dark with iron oxides.

The main question this poses for the silica cycle from 2.4 to 1.9 billion years ago is whether or not enough iron oxide was precipitating and enough BIFs were forming to be a globally significant output of silica from the ocean. Their formation appears to have been restricted to relatively shallow waters (perhaps because of their connection, via iron oxidation, to photosynthesis), and that represents a pretty limited extent compared to, say, the whole entire ocean.

At any rate, the cyanobacteria eventually screwed it up. Enough O_2 finally accrued from all their oxygenic photosynthesizing for ocean waters to become oxygenated. That meant they no longer held much dissolved iron. Which meant the end of widespread production of iron oxide particles. Which meant the end to the scavenging of silica. And that was the end of the formation of BIFs. Boo on you, oxygenically photosynthetically capable cyanobacteria.

9.9 And then Along Came True Silica Biomineralization

But eventually organisms like amoebae, choanoflagellates, sponges, and radiolarians started to biomineralize silica. As described in Chap. 5, this could have started somewhere between 770 million years ago and 550 million years ago. Almost certainly by 550 million years ago there was enough silica biomineralizing going on for the oceans to be no longer supersaturated with respect to silica. Silica-biomineralizing organisms were taking up dissolved silica and turning it into biogenic silica fast enough relative to dissolved silica's input via weathering to lower concentrations down below 1000 micromoles of silica per liter, although how far down below this point it is difficult to say.

We suspect that 550 million years ago and for some considerable time afterwards, the ocean supported relatively low amounts of biogenic silica production but, because dissolved silica concentrations were probably still relatively high, most of this biogenic silica became buried in ocean sediments rather than dissolving. This is quite different to the case today (high production, almost complete dissolution). The difference is that the early silica biomineralizers (sponges, radiolarians, and choanoflagellates) on their own would have been unlikely to produce as much biogenic silica per year as sponges, radiolarians, and choanoflagellates plus silicoflagellates and diatoms do today.

Diatoms, which only burst upon the scene about 150 million years ago, are mind-blowingly numerous and fast growing. Diatoms are also able to produce prodigious amounts of biogenic silica at near maximal rates even at vanishingly low dissolved silica concentrations and this is a somewhat unique feature of diatoms. They are the reason that today's ocean is deeply deficient in dissolved silica. Concentrations hovering around an average of 70 micromoles of silica per liter of ocean water must thus be a relatively recent phenomenon, one which dates back through less than the last 4% of Earth history.

In between the beginning of silica biomineralization and the first appearance of the diatoms, plants were learning to live on land and how to produce silica phytoliths and both of these things had huge effects on the cycling of silica. Invasion of the land and all the rooting of roots and excretion of carbon dioxide and organic acids that go with it increased the chemical and physical weathering rates of silicate minerals on land. This liberated significantly more reactive silica each year into soils (also a new invention), groundwaters, rivers, and, ultimately the ocean. Meanwhile the creation of phytoliths created the loop in the silica cycle that is the terrestrial biosphere that was discussed in Chap. 7. Over the last 440 million years since vascular land plants came into existence and more and more plants learned to make phytoliths, the terrestrial arm of the silica cycle grew ever larger and more dynamic.

The terrestrial silica cycle got a big boost with the appearance and evolutionary and ecological radiations of grass. Grasses, with their up to 10 dry weight percent silica in the form of phytoliths, popped up and grasslands began covering vast swaths of the continents starting at the end of the Cretaceous, that point in time also known as the last hurrah of the dinosaurs. That was only 60 million years ago. Imagine that. As recently at 60 million years ago, the land was carpeted in plants that weren't grasses and the amount of biogenic silica being produced on land was perhaps as little as a tenth of what it is today.

The latest great change to the silica cycle has been mankind and the changes we have wrought upon the surface of the planet. We've turned more than three quarters of the land surface over to intensive agriculture, buildings, parking lots, or paved roads, in the process diminishing silicate weathering rates and massively exporting silica from soils to sewage systems. We've also eutrophicated freshwater systems and built more than 850,000 dams that allow freshwater diatoms to bloom and then sink to the sediments, sequestering silica there, preventing it from reaching downstream areas and the ocean.

Because of this, the silica cycle right now is in a turbulent state. The amount of silica going in to ech reservoir of the silica cycle, be it soils, the biosphere, freshwaters, and the ocean and its sediments is not equal to the amount departing from those reservoirs. It will most likely take thousands of years for the silica cycle to settle down into whatever its new normal turns out to be.

Further Reading

Dürr HH, Meybeck M, Hartmann J, Laruelle GG, Roubeix V (2011) Global spatial distribution of natural riverine silica inputs to the coastal zone. Biogeosci 8:597–620

Fontorbe G, Frings PF, De La Rocha CL, Hendry KR, Conley DJ (2016) A silicon depleted North Atlantic since the Palaeogene: Evidence from sponge and radiolarian silicon isotopes. Earth Planet Sci Lett 453:67–77

Siever R (1992) The silica cycle in the Precambrian. Geochim Cosmochim Acta 56:3265–3272

Struyf E, Smis A, Van Damme S, Meire P, Conley DJ (2009) The global biogeochemical silicon cycle. Silicon 1:207–213

Tréguer P and De La Rocha CL (2013) The world ocean silica cycle. Annu Rev Mar Sci 5:477–501

Chapter 10
Silica Saves the Day

If pressed to sum up this book with two phrases, we'd pick *Silica is your friend* and *We owe just about everything to silicate weathering*. Not only does silicate weathering liberate silica from silicate minerals so it can participate in the silica cycle and all the biomineralization, diatom blooming, and collagen production that goes along with that, it has maintained the habitability of Earth continuously over the last at least 3.8 billion years. Take a moment. Let that thought sink in. Without silicate weathering, the surface of the Earth would be inhospitable. Uninhabitable. Totally lacking the biosphere that has needed at least 3.8 billion climatically reasonable years to build up in mass, complexity, and diversity to be what it is today (glorious, incredible, and taken for granted by its cleverest denizens). How does silicate weathering manage it? Read on to learn how silicate weathering acts as a thermostat that, within a wide range of solar luminosity (today's Sun being more than one-third brighter than its younger self that shone down upon the early Earth), has held Earth's climate within reasonable bounds for at least as long as there has been life on Earth. What's more, you'll hear how silicate weathering will mop up the mess we've made of climate with our overzealous emission of greenhouse gases into the atmosphere.

Now wait. Don't get too excited about this. Left to its own devices, silicate weathering will require 200,000 years to return the system to its natural state.

But, wait, maybe you can get a little bit excited about this. There is a possibility that we could artificially enhance silicate weathering, speeding it up so that it sucks out some of the sting of climate change from the coming decades. Unlike the next few hundreds of millennia, the next few decades are something that will be experienced by you, your children, and your grandchildren.

Go silicate weathering!

10.1 The Goldilocks Zone

If you've ever spent time with astronomers, astrobiologists, would be terraformers, or serious science fiction fans, you've heard someone yatter on about the Goldilocks Zone. The other name it goes by, the circumstellar habitable zone, gives the game away. The Goldilocks Zone is the range of orbits around a star that are potentially just right for the planet to be inhabitable. They're orbits that are not too close, and not too far away.

The idea of the Goldilocks Zone rests on the thought that although a lot of boxes need to be ticked for a planet to be feasible for life, there are some basic basics that are controlled to some degree by a planet's distance from its star (and upon the star's luminosity). The two basic basics are somewhat intertwined. Life, which to our understanding consists of self-reproducing, energy-gathering, energy-expending, membrane-bound containers carrying out chemical reactions in aqueous solution, needs liquid water. It also requires an atmosphere neither too crushing nor too vacuous. In other words, if it is to host life, a planet's surface should be neither too hot nor not too cold, and its atmosphere should not be too thick nor too thin. It all should be *just right*. Earth as we know it today, as you might imagine, is the type specimen of *just right*.

But of course it is but it isn't this simple. It isn't all about being in the perfect spot, even if this is a simple tale that gets told a lot. *Venus is too close to the Sun and that's why Venus is too hot. And Mars is too far away and that's why it's too cold. But Earth has just the right orbit, the perfect distance from the Sun, and so its climate is just right. We're so lucky!!*

But if you've ever cracked open a classic text on biogeochemistry, such as Wally Broecker's *How to Build a Habitable Planet*, you know that this isn't the entire truth. Yes, there are non-negotiable limits. A habitable planet can't be too close or too far from its star. But within that envelope, it's all about mitigating factors. Venus, for all its hellishness, lurks at the edge of the Goldilocks Zone. It actually isn't too close to the Sun for comfort. If Venus had not developed an insanely intense greenhouse atmosphere (which is thick enough to totally disqualify it on the crushing pressure front as well as the temperature front), the surface of Venus would be, at least on average and most especially in the shade, a chilly place. Meanwhile Mars could be warm enough for palm trees if it had more than a pittance of an atmosphere. Indeed, the terraforming of Mars, by beefing up its atmosphere with the right balance of oxygen, nitrogen, and greenhouse gases, into just such a pleasant state is the dream of many an armchair interplanetary explorer. And sharing Earth's just right orbit about the Sun (not to mention the Earth's rocky composition) is the Moon, who is no friend at all to reasonable temperatures, sufficient atmospheric pressure, or water in the liquid state (not to mention life).

No, to be a habitable planet takes more than an orbit within the Goldilocks Zone. Also mandatory is a greenhouse effect of the right strength and a natural, global-scale thermostat that is capable of keeping the greenhouse effect within reasonable bounds continuously over billions of years.

10.1 The Goldilocks Zone

So far, Earth is the only planet we know of that has succeeded on these fronts.

Providing greenhouse services, Earth has water vapor (its most powerful and abundant greenhouse gas), carbon dioxide (its most important greenhouse gas in terms of stabilizing climate), methane (brought to you in part by wetlands, termites, and burping cows), and several other natural (and now also some man-made and very powerful) greenhouse gases. For the thermostat, Earth has the weathering of silicate rocks.

10.2 Most of Us Can Model

To truly appreciate the climatic miracles worked by silicate weathering requires some backstory, namely, the explanation for why the surface of the Earth has the exact temperature that it does. So do you know what we're going to do? We're going to construct a simple model for average Earth surface temperatures. It will be based on the balance struck between the amount of sunlight coming in versus the amount of heat radiating back out to space. And all it will take to construct the model is some first principles from geometry and physics and a handful of measurements made by satellites.

Trigger warning: *there will be equations*. But you yourself personally won't have to do any math (unless you want to confirm the results). And, anyway, the math is the sort of math you learned how to do in 7th grade. Underneath the rust, it's still there. You can do it. You can be a climate modeler.

10.2.1 The Warmth of the Sun

Construction of our simple climate model begins with a very basic fact. For the surface of the Earth to be warmer than the space surrounding it (which is pretty damned cold), it needs a continuous input of energy. If we take a moment to think about the obvious, it's clear that the Earth's surface has two heat sources. One of these is the interior of the Earth and the other is the Sun.

The interior of the Earth is toasty warm, up to 900 °C (1600 °F) in the upper mantle and perhaps as much as 8000 °C (14,000 °F) toward the center of the core. As nature abhors a gradient, this heat continuously leaks out toward the cool surface of the Earth. If you've ever had a swim in a geothermal spring or ever soaked in a hydrothermal hot tub set up somewhere off the beaten path and underneath the stars or watched slowly flowing lava pour off into the steaming sea, you've enjoyed some of this heat, some of which is leftover from the planet's formation and some of which has been more recently generated by the decay of radioactive isotopes of elements like potassium, uranium, and thorium. At this point in Earth history, 43 terawatts is escaping to the Earth's surface through vents, fissures, springs, and volcanoes.

How much is 43 terawatts? It is 43 trillion watts, also known as 43×10^{12} joules of energy per second. If that still means nothing to you, then, in terms of power, 43 terawatts is the amount it would take to simultaneously run 700,000,000,000 sixty Watt light bulbs.[1]

Imagine the light. Imagine the heat. Imagine the electricity bill. And yet, in terms of heating the Earth's surface, the effect, on a global scale, is all but nil. Forty-three terawatts might be enough to move Earth's tectonic plates around and make mountains (and occasionally blow their tops off), but isn't enough to warm the entire surface of the planet by a noticeable amount.

Nope. To warm the surface of the Earth takes the 170,000 terajoules per second that reach the surface of the Earth from the great big round nuclear reactor in the sky. That reactor gives off heat and light created during the fusion of hydrogen atoms into helium atoms (it takes four hydrogens to make one helium, in case you were wondering). Turn off the Sun and the surface of the Earth would rapidly become a deep freeze.

As 43 terawatts is not even three hundredths of a percent of 170,000 terawatts, pffft. Why bother with it? And so our simple climate model is going to ignore the internal heat source. Our model only cares about the Sun.

So how bright is the Sun? And, just out of curiosity, over what kind of time scales does the brightness vary?

To answer the second question, solar luminosity, as it is more technically known, varies from day to day, year to year, and over eons. The hundreds of millions of year scale changes have to do with the Sun burning hotter as it ages.[2] The shorter scale variations (day to day, year to year) are related to fluctuations in the rate of delivery of hotter matter from deeper within the Sun toward its surface. We can now measure this shorter scale variability with sensors we've installed on satellites. But it's also possible to keep track of it by counting sunspots.

Sunspots deserve an aside even if they don't matter in the construction of a simple climate model where we will stick with the minimum number of features needed to explain the current average surface temperature of Earth (which is roughly 15 °C, or 60 °F). The solar variability behind sunspots would matter if we started using the model to understand changes in Earth's surface temperature in the past or if we tried to use the model to understand what might happen to climate in the future. But the real reason we'd like to explain to you why solar luminosity changes with the intensity of sunspot activity and in which direction is because sunspots are tricky devils.

---

[1] If you're old enough to remember what those are.

[2] As the Sun ages through its main sequence, it grows more luminous as helium, produced by the fusion of hydrogen atoms, accumulates in the Sun's core, increasing its density. The resulting contraction of the core increases the temperature of the Sun, which increases the rate of the nuclear reactions that produce helium, further increasing the density of the core and driving further increase in temperature. Thus, the Sun about four billion years ago was only 70% as luminous as the Sun of today.

10.2 Most of Us Can Model

Sunspots are more or less just that—darker than normal spots on the solar surface. But this darkness is a relative thing. Sunspots are most certainly still bright. They're just less bright than the solar surface surrounding them. The reason that they're less bright is that they're spots which are several hundred degrees cooler (and therefore much less luminous) than the normal solar surface.

And so you're thinking, *aha! When the Sun has a lot of sunspots, it must be dimmer than normal.* But it isn't true. When the Sun has a lot of sunspots, it's actually *brighter than normal.* (We told you they're tricky devils.) The reason for the Sun being brighter than normal when it has more sunspots than normal is that while sunspots are stagnant areas where not much hot, inner solar material is being convected to the surface, they're surrounded by extra super-hot, extra super bright areas of intense convection known as faculae. It's just hard for us to see extra super bright against a background of super bright, so we count the sunspots and not the faculae.

So if a lot of sunspots indicate a hotter than normal surface to the Sun, the opposite is also true—when there are fewer sunspots than normal, the Sun is slightly less luminous than normal. How much of a climatic effect high versus low sunspot activity can have depends on for how long the situation persists. Within the last 400 years for which we've been counting sunspots and writing the numbers down, there have been a few notable depressions in sunspot activity.

The Maunder Minimum, which lasted from 1645 to 1715, was spectacularly severe, indicating a decrease in the amount of sunlight reaching Earth of somewhere between 0.5 and 3 watts per square meter. Although this represented a drop of only 0.04–0.2% of the incoming solar energy, over the seven decades of the Maunder Minimum, like interest, it compounded. It is likely that the decreased solar luminosity during the Maunder Minimum in sunspot activity cooled climate, coinciding thus probably not coincidentally with the coldest years of the dramatic downturn in global temperatures known as the Little Ice Age.

The series of cold phases known as the Little Ice Age, while not a true ice age, ran from roughly 1300–1850. At its coldest, it was 2 °C colder than normal during the Little Ice Age, depressing crop yields and ushering in famine and plagues (including the Black Death), ice skaters on the Thames, and the severe winter that increased the misery and mortality of George Washington's troops at Valley Forge.

But, have no fear, at the moment, we're in the midst of a maximum in sunspot activity and the Sun's current total luminosity is 3.828×10^{26} watts. If you prefer, you can call that 383 yottawatts (and we'd love to say yow, that's a lotta yottawatts except that 383 probably isn't, although it is fair to say that one yottawatt is a lotta watts).[3] This brings us back to the task at hand-figuring out how much sunlight reaches the Earth.

Not all 383 yottawatts, of course. The Sun, a sphere, emits its energy in all directions, not just straight at Earth. As the energy radiates away from the Sun's surface, the spherical surface it is dispersed over grows bigger and bigger. By the

---

[3] Sorry.

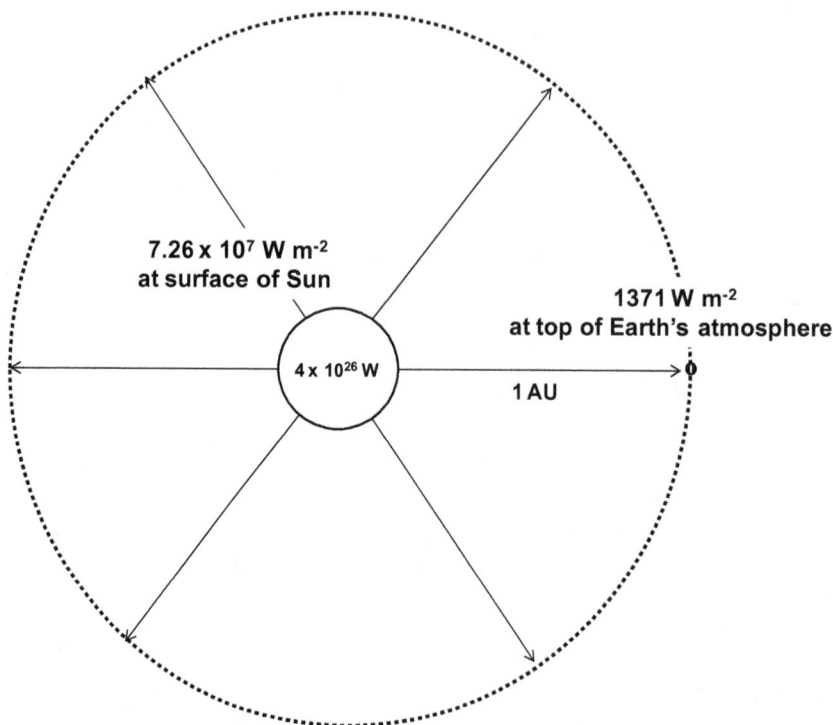

Fig. 10.1 By the time solar energy, radiating out in all directions from the Sun, reaches the Earth, it has been spread out over the surface of a sphere with a radius equal to the distance of the Earth from the Sun. On average this is 1 astronomical unit (plus, if you want to be as accurate as possible, the distance from the Sun's surface to its center). The figure is not drawn to scale!

time the spherical surface reaches Earth's average distance from the Sun, it has a radius of 93 million miles, or 1.496×10^{11} meters, as shown in Fig. 10.1.

Because the surface area of a sphere is equal to four times pi times the sphere's radius squared (that is, $4\pi r^2$), by this point, the initial 383 yottawatts emitted by the Sun has been spread over a spherical surface area of 2.8×10^{23} square meters. Dividing 383 yottawatts by 2.8×10^{23} square meters yields the density of power from the Sun by the time it reaches Earth: 1362 watts per square meter.[4] If our calculation is correct, that is what is reaching the top of Earth's atmosphere.

Theory is one thing and cold, hard measurements are another. So we've sent up satellites like Nimbus 7 and Earth's Radiation Budget Satellite (ERBS) to make measurements of the flux of solar energy reaching the Earth. How good was our

---

[4]In physics, this is known as the inverse power law which states that the intensity of light (I) at a given distance from its point source is equal to the initial power (P), in this case 3.828×10^{26} watts per square meter, divided by the area of the sphere with radius equal to the distance from the light ($4\pi r^2$). As an equation, this is $I = P/(4\pi r^2)$.

10.2 Most of Us Can Model

theoretical estimate? Celebratory Snoopy dance good. According to the satellites, the current and very badly named Solar Constant (that is, the amount of solar power reaching the Earth's upper atmosphere, something which is not a constant but a variable), is 1361 watts per square meter, give or take a watt or two. Not bad for a bit of middle school geometry.

From the Solar Constant, the radius of the Earth itself, and a few other geometric details, we can calculate that the total amount of solar power reaching the Earth should be that 170,000 terawatts that were mentioned earlier. This gives us enough information to make our first calculation of the surface temperature of Earth (and to test our simple model which currently says the only thing controlling Earth's average surface temperature is how much energy the surface of the Earth receives from the Sun).

If the temperature of the surface of the Earth is holding fairly steady,[5] we can say that Earth's average surface temperature is at steady state. It isn't heating up or cooling down fast enough for long enough for us to worry about in this calculation. When its temperature is at steady state, an object emits as much energy as it is receives (otherwise it would be warming or cooling). Here's where some basic laws of physics come in handy. The amount of energy emitted by an object at steady state can be predicted quite reliably from the surface temperature of the object. Which means that if we know how much energy is coming in from the Sun, we know how much energy is being lost to space (because it is the exact same amount as the energy coming in), and, all other things being ideal, we know how hot the surface of the Earth should be on average.

This is the basic law of physics known as the Stefan–Bolzmann Law. In mathematical terms, it looks like this:

$$E = \sigma T^4 \tag{10.1}$$

What that equation says is that the amount of energy (E) given off by an object (in units of watts per square meter of surface area) is equal to the object's surface temperature (T) in Kelvin raised to the fourth power times the Stefan–Bolzmann constant (notated as σ, the lowercase Greek letter sigma). The value of σ is 5.6×10^{-8} W m^{-2} K^{-4}, or, in plainer English, 0.000000056 watts per square meter per Kelvin raised to the fourth power. Kelvin, is just temperature in Celsius plus 273.

Because the amount of energy going in has to equal the amount going out, we can rearrange the equation and plug in the numbers to solve for T, the Earth's surface temperature.

$$T = \left(1361 \text{ W m}^{-2} / 5.6 \times 10^{-8} \text{ W m}^{-2} \text{ K}^{-4}\right) 0.25 = 395 \text{ K} \tag{10.2}$$

---

[5]This is a valid simplification to make. While the roughly 0.1 °C of warming per decade that has been occurring for the last hundred years is real, increasing, and climatically significant, it's not enough to do anything but get lost in rounding errors of calculations like these.

Again, to convert from Kelvin to Celsius simply requires subtracting 273. Thus according to the Stefan–Bolzmann Law, the Earth's surface should have an average temperature of 122 °C.

Oh.

To those of you who prefer Fahrenheit, that's 252 °F. Obviously not the right answer. This is hot enough to roast meat and if you take a moment to walk outside, even if you're in Death Valley in July, you'll be able to observe that the air at the Earth's surface is not hot enough to roast meat.

Hmm.

This leaves us with two choices. We could phone up all the physicists and tell them their Stefan–Bolzmann Law, which they've been using to deduce emissions from temperature and temperature from emissions since roundabouts the year 1880, is wrong. Or we could suspect that we've left something critical out of our model for Earth's surface temperature.

Conclusion: it isn't just solar luminosity that controls the surface temperature of Earth. We need to add in at least one other factor to our simple climate model.

10.2.2 Albedo, Which Is Not a Pasta Sauce

Anyone who has ever been snow blind could tell you what we've forgotten. The Earth can be pretty darned reflective. It does not absorb anywhere near 100% of the sunlight that reaches its surface. Pretty photos of a pretty bright Earth from space also give the game away. Earth reflects an awful of a lot of solar energy straight back out to space before it can do any warming of the planet at all. Shiny white things like ice and snow (and thick, glossy white paint) are especially good at this. Clouds, too, can be great at deflecting sunlight (although they can also be great at absorbing it). The Rolling Stones may not have been aiming to turn the Earth into a perfect absorber of sunlight, but that's what we'd end up with if they had their way. An Earth painted black (or more likely covered with even more paved roads, paved parking lots, and dark-roofed buildings), would reflect hardly any sunlight at all (provided you could also blacken the surface of the ocean).

All this reflectivity means that if we want to construct a more accurate heat budget from which to calculate Earth's surface temperature, we need to subtract what's reflected from the total amount of sunlight reaching the top of the atmosphere. But making an estimate of the Earth's reflectivity, something more technically referred to as its albedo, is a difficult thing to do. The texture and color of the Earth's surface is variable and always changing (snow falls then melts, plants wax and wane, asphalt grows less dark over time). On the other hand, the rocket scientists are all over it. Moderate-resolution imaging spectroradiometers (MODIS) in orbit on two of the satellites that make up NASA's Earth Observing System measure albedo all the time. The current typical number for Earth's average albedo

10.2 Most of Us Can Model

turns out to be 0.3 (on a scale from 0 to 1). The Earth, in other words, reflects about 30% of incoming solar radiation straight back out into space before it can warm the planet.

Thus the amount of energy that is coming in isn't 100% of the Solar Constant, but 70% of it. We need to adjust our simple climate model to take this into account. So instead of using the formulation we started with, the one where energy going out is related to the surface temperature, we're going to write an equation for the amount of energy coming in. That's still E, since E coming in equals E going out.

The amount of sunlight received is equal to the circular area of Earth that is sunlit (πr^2 where π is pi and r is the radius of the Earth) multiplied by the Solar Constant (SC) in watts per square meter. But this has to be also multiplied by the term $(1 - A)$ to take into account the reflection of some of the light. A is albedo, the current value of which is 0.3. Thus $(1 - A)$ is 0.7, meaning that only seventy percent of the sunlight is being absorbed. Thus, our equation for the amount of energy coming in looks like this:

$$E = (\pi r^2)(SC)(1-A) \tag{10.3}$$

Plugging in all the numbers gives us:

$$E = (1.278 \times 10^{14} \text{ m}^2)(1370 \text{ W m}^{-2})(0.7) = 1.2 \times 10^{17} \text{ W} \tag{10.4}$$

This 1.2×10^{17} watts, also known as 120,000 terawatts, is pretty much exactly seventy percent of the 170,000 terawatts that researchers have calculated (based on geometry) should be reaching the upper atmosphere. Consistency... hurrah!

Now to relate this to temperature. Energy in should be equal to energy out. And we have an equation for each of these (Eq. 10.1 for out and Eq. 10.3 for in). But, inconveniently, Eq. 10.1, the equation for the amount of energy being given off by Earth, is in units of watts while Eq. 10.3 is in units of watts per square meter. We can fix that by multiplying Eq. 10.1 by the spherical surface area of Earth (the term $4\pi r^2$) to put the loss of energy in terms of watts per square meter. Now we can set the flux of energy in and the flux of energy out as equals:

$$E_{in} = E_{out} \tag{10.5}$$

like this:

$$(\pi r^2)(SC)(1 - A) = (4\pi r^2)\sigma T^4 \tag{10.6}$$

The next thing we can do is simplify the equation by getting rid of terms that pop up on both sides of the equals sign and then dividing both sides by 4 (because we can). What we end up with is this:

$$(SC)(1-A)/4 = \sigma T^4 \tag{10.7}$$

This is starting to look like more than just an equation. It's starting to look like a bona fide energy budget model. Let's plug in some numbers to see if all it takes is solar luminosity and Earth's albedo to explain the average surface temperature of Earth.

In goes 1361 for SC, 0.3 for A, and the value of the Stefan–Bolzmann constant for σ. What comes out is a value for T of 256 K. That's -17 °C, also known as 1.4 °F.

Brr.

Another answer that can't be right. Earth's oceans are not solid ice from top to bottom.

And so we are faced with another two choices: give up and go back to the day job or wonder what else our simple climate model is missing.

Sorry if this seems to be a circuitous path. But what we're doing is creating and then testing hypotheses. So progresses science.

10.2.3 Emissivity

What is our previous calculation telling us? It is telling us that the Earth is losing heat as if its surface temperature was -17 °C rather than the roughly $+15$ °C that it is. In other words, the Earth is losing heat to space more slowly than you'd expect given how warm the surface of the Earth is. The most obvious explanation for this is that the Earth's atmosphere is insulated from the coldness of space.

This is where greenhouse gases come in.

And so we need to add one final term to our simple energy budget model for Earth's average surface temperature and we shall call it e. This e is for emissivity, or how perfectly emissive the Earth's atmosphere is. When e equals 1, the atmosphere is perfectly emissive; it entirely fails to impede the exit of energy to space. When e equals 0, the atmosphere is perfectly insulated; no heat is lost to space. Our model now looks like this:

$$(SC)(1-A)/4 = e\sigma T^4 \tag{10.8}$$

Emissivity is another slippery snake. It is what is decreasing when we add carbon dioxide and methane and other greenhouse gases to the atmosphere. But how much it is decreasing for every additional addition is a complex thing to figure out with great precision.

From a suite of satellite measurements and other data we can estimate that Earth's current emissivity is roughly 0.6 (although to do state-of-the-art climate modeling, we'd need to know the number to at least one more decimal place). The

10.2 Most of Us Can Model

final calculation to test the simple energy budget model for Earth's surface temperature then becomes:

$$T = ([[(3.8 \times 10^{26})(1-0.3)/4]/[(0.6)(5.6 \times 10^{-8})]])0.25 = 290 \text{ K} \quad (10.9)$$

If you subtract 273 from 290 K, you get 17 °C, or 63 °F, and that's a pretty realistic answer for the current average surface temperature of Earth.

Phew.

It took us a little while, but we got there. We closed the energy budget. It looks like our basic model is complete. We can now say with some degree of confidence that Earth's average surface temperature is controlled by three main variables: the Solar Constant (how much energy arrives at the upper atmosphere from the Sun), albedo (how reflective the Earth is), and emissivity (the greenhouse effect).

Nothing more, nothing less.

To cool the surface of the Earth you need to decrease the Solar Constant, increase the reflectivity of the Earth's surface, and/or decrease the greenhouse effect (which would increase emissivity). To warm the surface of the Earth, you need to increase the Solar Constant, decrease reflectivity, and/or increase the greenhouse effect.

Increasing the greenhouse effect is something we've been working on for some time now by adding carbon dioxide and methane to the atmosphere at what is a shockingly rapid rate. We're also indirectly decreasing reflectivity because the resulting warming is melting the sheets of ice on Greenland and Antarctica and shrinking mountain glaciers. That's a double-edged warming stroke, what the climate researcher refer to as a nonlinear effect (because one change doesn't bring about a simple, easily predictable amount of effect, but a set of effects that can amplify up to an enormous overall change in the system); adding greenhouse gases to the atmosphere doesn't just warm things by adding insulation, by melting continental ice, it also diminishes albedo, which results in even more warming of the climate.

These nonlinear effects are what makes it tricky (and so fascinating) to work on predicting the future, in terms of climate. First there is uncertainty in how much more carbon dioxide and methane (and other greenhouse gases) we will add to the atmosphere and how quickly. Then one has to figure out how much of that will end up not in the atmosphere, but in the ocean (the short answer is: less and less each year as the surface ocean saturates), and then how much the increase in the greenhouse insulation of the atmosphere and the resulting warming will cause changes in Earth's albedo.

So that's where climate researchers are today. Nobody's saying that greenhouse gases don't decrease Earth's emissivity. What we're trying to pin down is how rapid and how severe the changes to climate, the ice sheets, sea level, permafrost, forestation, agriculture, the pH of the surface ocean, ecosystems in general, economies, industries, regions, and nations are going to be.

10.3 The Importance of Greenhouse Gases

During this time of denial, responsibility avoidance, and slowly dawning realization that we may be in big trouble, greenhouse gases have gotten a bad rap. *Bad CO_2, bad CO_2.* And it is true that we've added (and are still adding) enough greenhouse gases to the atmosphere to decrease Earth's emissivity enough to have serious consequences for ourselves, our societies, our economies, and just about every creature in the biosphere. But greenhouse gases are also a major hero in the story of life's continuing existence on Earth.

It's really quite a slight that none of the ancient religions gave greenhouse gases godhood beside such things as Sun, Earth, sea, death, harvest, love, joy, wine, wisdom, and warfare. As our construction of a simple climate model has revealed, the Earth is so reflective that without greenhouse gases in its atmosphere, the Earth's surface temperature would be that uninhabitably chilly −17 °C (or worse, considering that if the Earth's surface was that of a snowball, its reflectivity would be through the roof).

The ease with which Earth's surface temperatures could be totally impossible for life poses a puzzle. Sunlight coming in, sunlight reflecting out, and the rate at which heat is trapped by the atmosphere have to be cumulatively *just right* for it all to work out. How on Earth does that happen? How has the Earth has managed to have a greenhouse gas content that is somehow *just right* for so many billions of years?

This question becomes even more interesting when you realize that Earth is the only planet in our Sun's Goldilocks Zone (or indeed any planet that we know of at all) to have managed such a feat. Even more amazingly, due to the Sun growing hotter as it ages, today's Sun is about 40% brighter than the Sun of four billion years ago, and yet climate has been reasonably reasonable for all that time. How? How? How? How?

To go back to Venus, this hottest planet in the solar system most certainly didn't manage to stay temperate. Its average surface temperature is 462 °C (864 °F). But given its reflectivity, its average surface temperature ought to be −46 °C (−51 °F). Yes, you read that right. Even though Venus is so close to the Sun, if it didn't have any greenhouse effect at all, it would be almost as cold, on average, as the Earth's South Pole.

But Venus has a tremendous greenhouse effect on two counts. First, its atmosphere is 96% carbon dioxide (our atmosphere, even with the fossil fuels we've burned so far, is only 0.04%). Second, that 96% is 96% of an atmosphere that is about 90 times thicker than Earth's. Put those two things together and you end up with enough insulation to boost the surface temperature of Venus by roughly 500 °C.

The same thing could have happened on Earth. We too could have ended up with a runaway greenhouse effect. It would have gone like this. A little bit of warming would have increased the evaporation of water out of the oceans and off of the land, increasing the amount of water vapor in the atmosphere. Water vapor happens to be a potent greenhouse gas (most of Earth's greenhouse effect is actually due to water vapor, not to carbon dioxide). The increased greenhouse effect would have

10.3 The Importance of Greenhouse Gases

increased the surface temperature of Earth. This would have evaporated even more water into the atmosphere. Which would have further increased the greenhouse effect. And so on, until all the water was in the atmosphere and the surface of the Earth was hot enough to boil off rock. And that would have been totally game over for life on Earth.

But it hasn't happened, either ever or not since the end of the Late Heavy Bombardment phase early in Earth's history discussed in Chap. 2.

Why hasn't it happened? What has prevented Earth's greenhouse effect from running up and out of control?

Silica, of course. In the form of silicate weathering.

10.4 Silicate Weathering Consumes Carbon Dioxide

We've mentioned this so many times, by now you might be sick of it. The weathering if silicate minerals converts carbon dioxide, a powerful greenhouse gas, into solutes such as carbonate and bicarbonate ion as exemplified by the idealized silicate weathering equation, that for the dissolution of the silicate mineral wollastonite ($CaSiO_3$) into solutes such as calcium ions (Ca^{2+}), dissolved silica ($Si(OH)_4$), and carbonate ions (CO_3^-):

$$CaSiO_3 + 2(H_2O) + CO_2 \rightarrow Ca^{2+} + Si(OH)_4 + CO_3^- \qquad (10.10)$$

The role of carbon dioxide in this and other silicate weathering reactions is to react with water to form carbonic acid (H_2CO_3) which can provide the protons (H^+ ions) necessary for the dissolution reaction. As the mineral dissolves, liberating silica tetrahedra (SiO_4^{4-}), the tetrahedra steal the protons to form dissolved silica ($Si(OH)_4$). This leaves what was formerly carbon dioxide stuck with the oxygen atoms from water to become carbonate ion (CO_3^-) or, if the water is still acidic enough to have protons to spare, bicarbonate ion (HCO_3^-).

There are two reasons why it's so darned important that silicate weathering converts carbon dioxide into carbonate and bicarbonate ions. The first is that if there wasn't a process to remove carbon dioxide from the atmosphere of Earth, carbon dioxide would have long ago built up to runaway greenhouse effect levels. Like heat, carbon dioxide is constantly leaking out of the interior of the Earth, where it exists in abundance, via geothermal springs, hydrothermal vents, fissures, and volcanoes. The current rate of such mantle outgassing of carbon dioxide, as it is known, is 150 megatons (aka 150 million tons) of carbon each year. If there was not a process removing this carbon dioxide as fast as it is being added, it would only take a million years for the atmosphere to be packed with 150,000 gigatons of carbon—an atmospheric content of carbon dioxide of 7%. That would be utterly climatically catastrophic. (Just for comparison, today we're entirely reasonably panicking about having hit 0.04%.) The resulting acidification of ocean, lake, stream, and river waters would be incredibly devastating, too.

But, plants, you might be thinking. Surely they will take up all that carbon dioxide and stop it from over-accumulating. But, despite their prodigious production of organic matter via photosynthesis, plants don't prevent such a build-up carbon dioxide. Except during brief and exceptional moments in geologic history, such as when evolutionary innovations allowed animals and plants to invade the land and fill it up, the biosphere does not notably increase in mass. It is, essentially, at steady state most of the time. The amount of carbon that is fixed into the biosphere via photosynthesis each year is pretty close to equal the amount that is breathed out of it in the form of carbon dioxide. Okay, yes, there is some permanent sequestration of carbon into organic matter in sediments, but it's nowhere near 150 megatons of carbon a year. (And, if anything, because of deforestation, especially by burning, right now we're *decreasing* the amount of biomass in the biosphere, sending all that carbon dioxide into the atmosphere.)

It mainly falls on silicate weathering to convert the carbon dioxide in the ecosphere[6] into the carbonate and bicarbonate ions. In the shorter term (timescales of thousands of years), this helps by allowing the ocean to store the bulk of the reactive carbon in the ecosphere in the ocean. In the longer term, it allows the carbon to be removed as calcium carbonate minerals.

The numbers on this are clear. Of the 40,000 billion metric tons (aka gigatons) of carbon dispersed through the ecosphere, 37,000 billion metric tons are stored in the ocean mainly as carbonate and bicarbonate ions. Living creatures and no-longer-living organic matter in the ecosphere comprise 2000 billion metric tons of the remainder, leaving only 700 billion metric tons of carbon to exist in the atmosphere as carbon dioxide. So already without silicate weathering's production of carbonate alkalinity, we'd be in big trouble. Nobody wants to try to live on an Earth where the greater part of 40,000 gigatons of carbon resides in the atmosphere as carbon dioxide.

But the carbonate and bicarbonate ions produced from carbon dioxide during silicate weathering don't just wash out to the ocean and accumulate forever. Instead, carbonate and bicarbonate ions eventually become incorporated into carbonate minerals. Today mostly this means the calcium carbonate ($CaCO_3$) of the scales produced by calcareous coccolithophorids, of the chambered shell of a foraminiferan, of the rather more recognizably snail-like shell of a pteropod, or of the skeleton produced by a coral. These carbonate minerals then end up in ocean sediments. As with the biogenic silica that is buried within ocean sediments, the calcareous sediments ride the plate tectonic railway back down into the mantle where they melt and recycle their rock-building and carbon dioxide-forming constituents back into the upper interior of the Earth.

Thus no matter how much all those volcanoes burp, on geologic time scales, carbon dioxide isn't building up to astronomical levels within Earth's atmosphere and Earth is not having to face the prospect of warming until it boils off its own oceans. Instead there is this long-term carbon cycle that cycles carbon through the

---

[6]Earth's ecosphere consists of the atmosphere, oceans, and land and the life within or upon them.

10.4 Silicate Weathering Consumes Carbon Dioxide 191

mantle, out into the ecosphere, and then back down into the mantle. This long-term carbon cycle is intertwined with the long-term silica cycle because the volcanism which helps to return carbon dioxide to the atmosphere also forms silicate rocks and because the weathering of silicate rocks converts carbon dioxide to carbonate and bicarbonate ions that will become carbonate sediments that, along with silica sediments, return to the mantle via subduction and melt, releasing carbon dioxide and the raw materials needed to make new silicate rocks.

So far so amazing. But it begs one final question. How is it that silicate weathering removes just the right amount of carbon dioxide from the atmosphere so that the greenhouse effect of Earth is just right, holding climate within reasonable bounds?

10.5 The Temperature Dependence of Silicate Weathering

To consume carbon dioxide is one thing, to regulate it to hold average planetary surface temperatures within a moderate range is, by orders of magnitude of amazingness, another. Accordingly, how silicate weathering pulls this off so successfully has been a million dollar question for quite some time (about 75 years… an eon or two in terms of the onward march of science). Already by 1952, Harold Urey (known for enriching uranium via gaseous diffusion, collaborating on the famous Miller-Urey experiment, and pioneering the use of stable oxygen isotopes in the reconstruction of paleoenvironmental conditions, all after winning a Nobel Prize for the isolation of deuterium) had noted in scientific publications and seminars that silicate weathering consumes carbon dioxide and that this is what prevents all the carbon dioxide from the mantle building up in the atmosphere. But it was not until 1981 that a trio of planetologically inclined scientists—James Walker and Paul Hays of the University of Michigan and Jim Kasting, then at the National Center for Atmospheric Research in Colorado—hit upon the key reason why silicate weathering is the key reason for Earth's long-term balminess. The only way that this can be, they pointed out, is if to a very large degree the speed at which the chemical weathering of silicate minerals occurs depends on climate temperature.

It is easy to imagine the stunned silence, the hands slapped on foreheads, the crushing feelings of *Why didn't I think of that?* as researchers read the paper or otherwise got wind of its proposition. All the best ideas, and this is most certainly one of them, seem obvious in retrospect and it is impossible to understand how it is that nobody thought of them before.

The way temperature's control of weathering and weathering's control of temperature works goes like this: When the surface of the Earth is colder than normal, the cold slows down the overall rate of silicate weathering. But the rate at which the mantle is outgassing carbon dioxide to the atmosphere is pretty much unaffected by climate. The amount of carbon dioxide consumed by silicate weathering each year is less than the amount being added to the atmosphere from the mantle. It's not a huge imbalance, most likely far less than one percent. But it is enough of an imbalance that slowly, over tens of thousands of years, atmospheric concentrations of carbon dioxide begin to increase.

The increase in carbon dioxide concentrations is a subtle one, enormously less impressive than the skyrocketing going on today due to our burning of fossil fuels and forests. But it is enough to progressively increase the greenhouse effect, slightly slowing down the Earth's loss of heat to space. Slowly, very slowly, the surface of the Earth begins to warm.

As global climate warms, silicate minerals begin to weather more quickly, increasing the rate at which carbon dioxide is converted to carbonate and bicarbonate ions. Eventually silicate weathering is running fast enough for the removal of carbon dioxide from the atmosphere each year to exceed its input from the mantle. The imbalance has tipped to the other direction. Atmospheric concentrations of carbon dioxide begin to decrease.

As atmospheric carbon dioxide concentrations decrease, so does the greenhouse effect. Earth's surface temperatures begin to decrease and as they do so, global rates of silicate weathering also decrease. Eventually they are low enough that atmospheric carbon dioxide begins to build back up again.

And so it goes, on and on and on through the eons, containing Earth's total greenhouse effect within the realm of reasonable.

This... *this* is how you keep a planet habitable for billions of years. Isn't it brilliant? GO SILICA.

Of course there are details to deal with, most of them making changes to Earth's reflectivity or the strength of its greenhouse effect or to the Solar Constant much more rapidly than silicate weathering can adjust the amount of carbon dioxide in the atmosphere in compensation. Continental ice sheets can grow or collapse. Permafrost can melt, liberating prodigious quantities of methane. Sunspot activity can shift into higher or lower gear. The list is long and it leaves lots of climatic wiggle room even within a climate system that is still ultimately stable. In addition, the topography, latitudinal distribution, and mineralogy of continents, the seasonal and geographic distribution of rainfall and its intensity, the physical and chemical activities of plants (especially their roots, down in those soils where a lot of weathering occurs), the physical and chemical activities of animals (especially modern humans), and many other factors change the relationship between climate temperature and silicate weathering rate over time such that the global average silicate weathering rate, over geologic time, is not a steady and perfectly predictable function of global average surface temperature. Altogether this makes it possible to sometimes have a warm "greenhouse" Earth totally lacking in continental ice sheets (for example, during dinosaur times) and at other times a chilly "icehouse" Earth (for example, now), where thick sheets of ice cover large areas of continent while the basic tenant remains true—silicate weathering rates are to a first order temperature dependent and that's why Earth, which still has water and a tectonic cycle, is the only planet in the Goldilocks Zone that is actually quite seriously inhabitable.

10.5 The Temperature Dependence of Silicate Weathering

Of course there have been vicious arguments about the details. With our little ape brains, it's hard to imagine all the interacting mechanisms by which this regulation of climate by silicate weathering can be simultaneously loose but (so far) foolproof.

For instance, it would be only a slight exaggeration to say that fist-fights have broken out over whether India plate tectonically ramming Asia to create the super steep, highly physically weatherable Himalayas and the super warm, wet, chemical weather friendly South Asian monsoon has reset the silicate weathering thermostat. Spraying so much warm water on steep, easily weatherable terrain could make the silicate minerals there so easy to weather, goes the thought, that higher than normal global average silicate weathering rates could be sustained at lower than normal global average temperatures. This could be what kicked off and helped to sustain the more or less continuous cooling of Earth's climate over the last 55 million years, the one that took us from the warmth of the dinosaur world to the iciness of the ice ages. It's a seductive, simple story that is certainly too good to be true but also isn't entirely false, which is probably why it has had so many fiendishly fierce detractors as defenders over the last 20 years. Uplift of the Himalayas may have helped, but if they did so, their actions added to what was going on with the arrangements of the continents and the resulting changes to ocean circulation that allowed the poles to cool enough to form ice sheets that were able to collect upon Antarctica and then, later, Greenland, Northern Europe, and northern North America.

10.6 The Paleocene-Eocene Thermal Maximum

But sometimes it happens that silicate weathering's climate moderating skills shine utterly, perfectly clear. Take the Paleocene-Eocene Thermal Maximum, or PETM for short. It occurred 55 million years ago, at the boundary between those two geologic epochs, the Paleocene and the Eocene, and was a time when silicate weathering shooed global temperatures back down to where they should have been following a greenhouse gas catastrophe of equal (or perhaps somewhat lesser) severity to the one we are creating.

The PETM began with the release, over a few thousand years, of a huge amount of carbon to the atmosphere in the form of a mixture of carbon dioxide and methane, both of which are greenhouse gases. This release resulted in massive global warming and the bad things, like the widespread extinction of species, that go along with it. But the final 200,000 years of the PETM consisted of the slow, steady removal of carbon from the system due to increased rates of silicate weathering. By the time the PETM was over, climate was back to where it would have been and everything was hunky dory (unless you were one of the many species that was driven to extinction in the interim).

To fans of silicate weathering, the PETM rocks! It's a natural, global-scale experiment on greenhouse gases, silicate weathering, and climate, long since run, whose results are recorded in marine sediments.

The PETM began in the North Atlantic because, well, it was caused by the beginning of the North Atlantic. Between 300 million years ago and about 180 million years ago, the Earth only had one continent. Its name was Pangaea and it consisted of pretty much all the land you can imagine today all glommed together into one colossal land mass (check out the upper left panel of Fig. 10.2).

But, as we heard in the previous chapter, even a supercontinent does not last forever. At some point the convection of hot mantle up underneath Pangaea in many places broke it into segments (namely, more or less the continental masses we know today) that were then shuttled apart.

This reconfiguration was long and drawn out, as shown in the snapshots of Fig. 10.2. The lower latitude North Atlantic began to form about 200 million years ago, separating what is now North America from what is now Africa, but the northern North Atlantic, that which now lurks in between Canada, Greenland, and

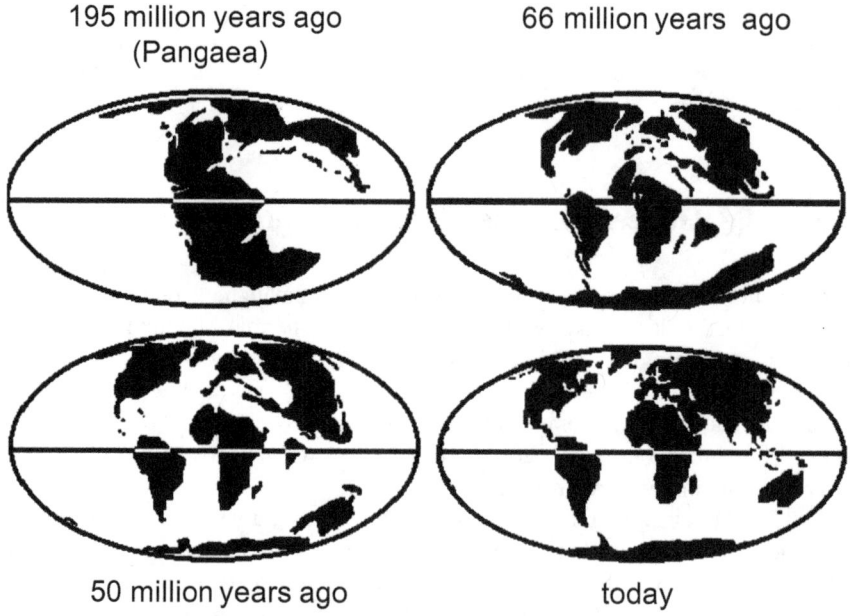

Fig. 10.2 The continents through time. In the *upper left* is the supercontinent Pangaea as it looked 195 million years ago. In the *upper right* shows the breakup of Pangaea's progression by 66 million years ago (about 10 million years before the PETM). The *lower left* shows how far things had progressed by 50 million years ago, 5 million years after the onset of the PETM (note how much further from Europe Greenland now is). Today's continental configuration is shown in the *lower right*. These images are based on those of the plate tectonic maps and continental drift animations by CR Scotese, PALEOMAP Project (www.scotese.com) and included here with permission

10.6 The Paleocene-Eocene Thermal Maximum

Europe, waited until 56 million years ago to begin opening up, creating the northern portion of the North Atlantic ocean basin.

It was the burst of volcanism that started separating North America from Greenland from Eurasia that spewed intense amounts of carbon dioxide into the atmosphere from the mantle in a relatively short period of time. That was bad enough on its own. But what was worse was this volcanism ran into some carbon-rich rocks or sediments present on the continent. All that magma hitting all these carbon-rich rocks volatilized them, sending even more voluminous additional quantities of carbon dioxide (and also methane, an even more potent greenhouse gas) into the atmosphere in a short period of time.

All in all, over approximately 4000 years, at the rate of perhaps just one gigaton of carbon per year (which is about one tenth of the rate at which we are currently releasing carbon dioxide by burning fossil fuels), somewhere between 2000 gigatons of carbon and 4500 gigatons of carbon was vented to the atmosphere. Because of the release of carbon dioxide and methane (which within a few years oxidizes to become carbon dioxide) over several thousand years, concentrations of carbon dioxide in Earth's atmosphere increased by something between 500 parts per million and 2500 parts per million.

Astounding. Incredible. And unprecedented. (Until now. At our current rate of 10 gigatons per year, we'll only need 200–400 years to match the total PETM carbon release.)

The predictable result of all the carbon added to the late Paleocene atmosphere was a nasty increase in the greenhouse effect. As we can see from the geologic record from all over the Earth, from the middle of continents to the bottom of the sea, things got hot, hot, HOT. Temperatures across the globe (including in the deep ocean), which were already warm compared to today, soared by an additional 5 to 8 Celsius degrees (roughly 9 to 18 Fahrenheit degrees) and stayed that way for tens of thousands of years. This was extreme. Many marine and terrestrial organisms had to pick up sticks and move great distances to find still tolerable climes or die (and die they did in droves). Storms got stormier but at the same time, large tracts of land dried out because the periods of time between storms grew longer. Did we mention that organisms (and entire species) died in droves?

But eventually all that extra volcanic energy and all that extra vaporizable carbon exhausted themselves. The Earth's release of carbon dioxide and methane to its atmosphere quieted back down to more normal levels, for example, for carbon dioxide of about 150 megatons per year. The climate crisis could stop getting worse, but there was still that total of 2000 to 4500 extra gigatons of carbon stuffed into the ecosphere and wreaking global warming.

But here's where silica charges into save the day. From day one of the warming of climate, rates of silicate weathering on the continents had increased. As there are both physical and chemical limits, silicate weathering rates couldn't ever be rapid enough to keep pace with the release of one gigaton of carbon per year during those first years of the PETM. But once the excessive release of carbon into the atmosphere ceased, the elevated silicate weathering rates could remove carbon dioxide faster than it was being put in. It took that 200,000 years to fully accomplish, but the excessive excess of carbon dioxide was converted into carbonate ion and bicarbonate ion that were ultimately turned into calcium carbonate by foraminiferans, coccolithophorids, mollusks, and corals and deposited on the seabed as sediment. Thanks to the silicate weathering stimulated by the heat of the PETM, the extra 2000 to 4500 gigatons of carbon that were heating the surface of the Earth via the greenhouse effect and horrifically acidifying the ocean had been removed from the ocean-atmosphere system.

If we drill down into the seafloor and remove cores of sediment from this time and measure the stable isotopic composition of the microscopic calcareous tests (shells) of benthic (bottom-dwelling) foraminiferans, we can see that this happened (see Figs. 10.3 and 10.4). The rapid drop in the ratio of carbon-13 to carbon-12 in the foraminiferan tests, which is due to a change of the ratio in the carbon dioxide dissolved in seawater, reveals that the massive input of carbon to the ecosphere at

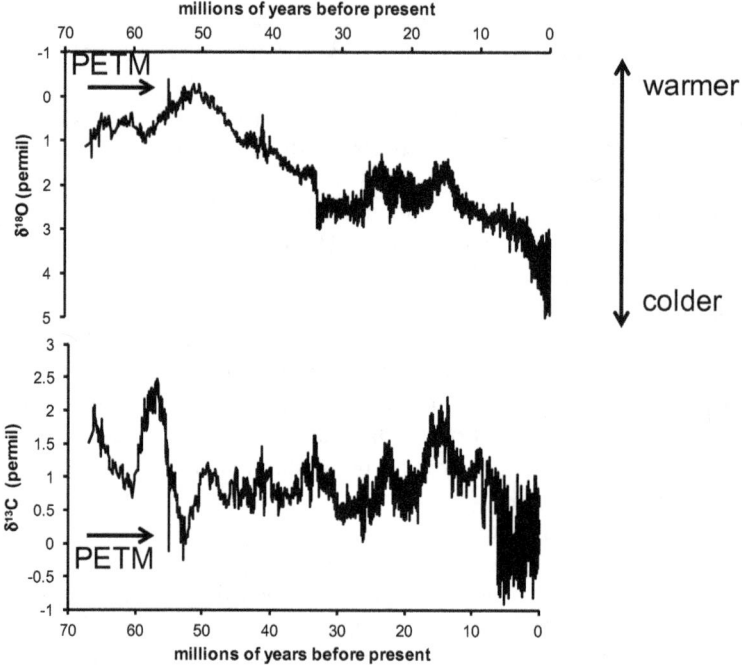

Fig. 10.3 The long cooling of climate during the last 55 million years as revealed by the oxygen isotopic composition of the calcium carbonate tests of bottom-of-the-ocean-dwelling foraminiferans (*upper panel*). The higher the value for $\delta^{18}O$ (a measure of the ratio of oxygen-18 to oxygen-16), the colder the temperature of the bottom of the ocean when the foraminiferan biomineralized its test (shell). Note that $\delta^{18}O$ values are plotted in reverse (highest to lowest) so as to give a direct sense of the change in temperature. The PETM is visible as the spike in values 55 million years ago indicated by the *arrow*. The *lower panel* shows the stable isotopic composition ($\delta^{13}C$) of the same material. The sudden shift in $\delta^{13}C$ at the PETM is also visible as a spike in values and is also indicated by an *arrow*. These figures have been drawn using data from Zachos et al. (2001) Science 292, 686–693

10.6 The Paleocene-Eocene Thermal Maximum

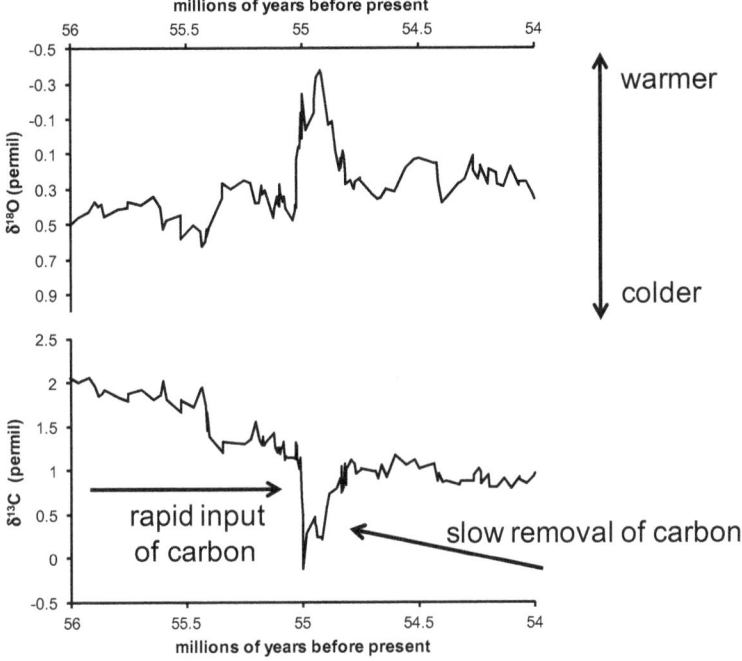

Fig. 10.4 This figure presents the same data as Fig. 10.3, but zoomed in on the time periods of 56–54 million years ago. The isotopic excursions of the PETM are clearly visible

the beginning of the PETM is matched by a massive drop in the oxygen-18 to oxygen-16 ratio that indicates the soaring of temperature (note that oxygen isotopes are plotted in reverse on the figures, such that down is up).

The following slow shift of these ratios back toward "normal" is what reveals that the added carbon, which had a notably low carbon-13 to carbon-12 ratio, was being removed from the ecosphere (via deposition, following its conversion to carbonate and bicarbonate ion, as calcareous sediment that will all ultimately become subducted back down into the mantle). At the same time, the increasing stable oxygen isotope ratios indicate temperatures slowly also returning toward "normal" for that moment in Earth history.

We also know from a shift in certain trace elements and isotopes of other elements in seawater (that were incorporated into the calcium carbonate of the foraminiferans who lived in the seawater) that silicate weathering rates had increased during this time. So while the changes in the carbon and oxygen isotopes are the records that show the rise then drop in temperature and the addition then removal of the excess carbon, the other records are the smoking gun that silicate weathering was the hero of the story. In other words, in the face of temperatures soaring up to 8 Celsius degrees above normal, silicate weathering rates significantly increased, just as we have hypothesized they should. Thus, the initial step in carbon dioxide removal was indeed its conversion into carbonate alkalinity. The second step of the process—the production and sedimentation of calcareous biominerals from the carbonate and bicarbonate ions added to the ocean by silicate weathering, can also be seen in the worldwide increase in carbonate sedimentation rates at this time, and in an increase in ocean alkalinity (visible in the patterns of carbonate sediment accumulation within the ocean).

Altogether, the paleoceanographic records (and there are hundreds of them) from the PETM tell a convincing story of the danger of pumping too much carbon dioxide into the atmosphere and of the slow power of silicate weathering to save the day.

10.7 Enhanced Weathering

Another reason that the PETM is so interesting is that it holds lessons for us regarding the situation we're creating today. You can't look at the records that suggest a massive increase in the amount of carbon dioxide to the atmosphere was followed by significant global warming and not think that we're on track to totally screw up climate in ways that will be huge, global, and devastating. At the same time, it's heartening to see that silicate weathering will eventually mop up the mess, although eventually means over the course of 200,000 years.

We won't be around for that and neither will 10,000 generations of our descendants. Our entire species may not even last that long just in general and there are many, many other species facing extinction for sure within the next few decades due to climate change and other stressors. Naturally speaking, silicate weathering offers us no comfort from the disaster of our own creation.

But we're human beings. We love to tinker. We get stuff done on grand scales. That's our strong point (and our curse). Surely we can do better that wait for silicate weathering to chug through the problem. Surely we should be able to, for instance, figure out how to extract excess carbon dioxide from the atmosphere.

People are working on it. But removing the carbon dioxide we've added is tricky. There's enough up there to cause a huge problem, climatically speaking, but that amount is still only 400 parts per million, or 0.04% of what's in the air. It's hard to extract something that exists at such a low concentration. Right now, the only things that can efficiently suck out carbon dioxide at such low concentration in air are plants and photosynthetic microorganisms and it took them hundreds of millions of years of evolution to be able to do it well.

And even if we could start yanking massive amounts of carbon dioxide out of the atmosphere and concentrating it somewhere, where would we put it? Even compressed to great pressure to become a difficult and dangerous thing to contain, that amount of carbon dioxide would take up an immense amount of space. The technical and practical challenges to carbon storage are imposing.

10.7 Enhanced Weathering

But what if instead of extracting and storing carbon dioxide we could speed up the rate of silicate weathering, the natural process by which all of this excess carbon dioxide will be removed from the atmosphere? What if we could, for instance, increase global silicate weathering rates by a factor of ten? Would that be enough to avoid dangerous amounts of climate change? The bonus about enhancing weathering is that not only are there no carbon storage issues, the carbonate and bicarbonate ions added to the ocean would start counteracting the horrific acidification of the ocean we've set off by spewing so much carbon dioxide (which, when dissolved in water, forms carbonic acid).

Here is how enhancing silicate weathering rates would work. First prodigious quantities of an easily weatherable silicate mineral, such as the green olivine that, for instance, makes up the green sand of famed Papakōlea Beach in Hawaii, would need to be mined each year and ground into a fine powder. The powder would then be transported to warm, wet tropical regions and, ideally, spread over farmland (although grasslands, forest, and jungle would also do). Each batch of olivine powder would need only a couple of years to dissolve, in so doing, converting a bit of atmospheric carbon dioxide into carbonate alkalinity that would ultimately end up in the ocean. At the same time, dissolution of the olivine would release nutrients, such as dissolved silica and also phosphorus and potassium (which are trace inclusions in the mineral) to the soils, boosting plant growth and agricultural yields. A win-win-win, in principle.

What would it take to pull it off at a scale large enough to matter? First and foremost, people would need to agree to it, especially the people living in the humid tropical localities where the olivine powder would be deployed (because warm and humid is where the olivine would weather the fastest and most completely). (In other words, the people who caused the least amount of the problem of excess greenhouse gases would have to be the people who literally live with the attempted solution; we'd probably have to make it more worth their while than it would be on its own.) Second, carrying out such geoengineering of climate on a grand scale would require a lot of investment in infrastructure and a sufficient collection of governments, foundations, and private enterprises willing to spend the money to do it. Thousands of miles of new (or improved) roads would be needed for the efficient delivery of olivine powder to the fields where it would be applied. Facilities capable of grinding tons and tons of olivine powder would need to be built in numerous, strategic locations. Olivine mining efforts would have to be expanded.

Those are all tall orders. But if we could make it happen, enhanced weathering, deployed to its maximal and most efficient extent, could remove one billion tons of carbon from the atmosphere each year.

One billion tons. Is that enough?

As amounts go, this one billion tons is both a lot and, on its own, not enough.

Currently we're releasing 10 billion tons of carbon each year by burning fossil fuels and converting forests, jungles, and grasslands to farmable fields, etc. Overall, that means we could compensate for only 10% of our yearly emissions (as long as they don't, *ahem*, increase).

But it is also fair to point out that not all 10 billion tons of carbon that we release ends up in the atmosphere, "only" 3.5 billion tons does. The rest of it gets soaked up by the ocean, acidifying it, and some of the carbon dioxide goes into re-growing previously destroyed forests. So, at least until the ocean and land sinks saturate and stop soaking up some of our carbon dioxide, enhanced weathering could decrease the annual increase in atmospheric carbon dioxide by about a third and that's nothing to sneeze at.

This illustrates a point that just about everyone who studies how to fight global warming inevitably comes to realize. There is no one technology that can solve the problem for us. We can't just keep releasing massive amounts of carbon dioxide to the atmosphere in the blithe belief that before the effects go catastrophic, we'll figure out one simple way to pump it all back out. Our rates of carbon dioxide emission are long past the point where a geoengineering technique like enhanced weathering which would speed up the natural processes that right the system could match them. But if we deployed it while lowering our rate of carbon dioxide emissions,[7] we could possibly avoid catastrophic amounts of global warming.[8] Otherwise, it's like the PETM all over again. Natural plus enhanced silicate weathering won't be able to keep pace with the addition of carbon dioxide to the atmosphere and global temperatures will skyrocket and not come all the way down for 200,000 years.

We could end this book by telling you that in the human race, silica has met its match and been bested. But, you know, even if on its own it can't stop us from (temporarily) screwing up the ecosphere, silicate weathering will be going on long after our species has gone extinct. And, ultimately, silica will still save the day, removing our greedily emitted carbon dioxide out of the atmosphere and into the ocean, down into the sediments, and then further down into the mantle. We just won't last long enough to witness it.

Further Reading

Alley RB (2016) A heated mirror for future climate. Science 352:151–152
Broecker WS (1985) How to Build a Habitable Planet. Eldigio Press, Palisades
Hartmann J, West AJ, Renforth P, Köhler P, De La Rocha CL, Wolf-Gladrow DA, Dürr HH, Scheffran J (2013) Enhanced chemical weathering as a geoengineering strategy to reduce atmospheric carbon dioxide, supply nutrients, and mitigate ocean acidification. Rev Geophys 51:113–149

---

[7]The way for a planet of 7 billion human beings to do that not catastrophically is by building up the technology and infrastructure needed to power our lives using wind, solar, and other low- to no-carbon means.

[8]The key word here is possibly. We have already set a staggering amount in motion, enough that summer sea ice in the Arctic Ocean will within a handful of years be a thing of the past, for instance, and all the side effects that go along with that.

Further Reading

Langmuir CH, Broecker W (2012) How to Build a Habitable Planet: The Story of Earth from the Big Bang to Humankind. Princeton University Press, Princeton

Walker JCG, Hays PB, Kasting JF (1981) A negative feedback mechanism for the long-term stabilization of Earth's surface temperature. J Geophys Res 86:9116–9782

Zachos J, Pagani M, Sloan L, Thomas E, Billups K (2001) Trends, rhythms, and aberrations in global climate 65 Ma to present. Science 292:686–693